Water Chlorination/ Chloramination Practices and Principles

AWWA MANUAL M20

Second Edition

American Water Works Association

Science and Technology

AWWA unites the entire water community by developing and distributing authoritative scientific and technological knowledge. Through its members, AWWA develops industry standards for products and processes that advance public health and safety. AWWA also provides quality improvement programs for water and wastewater utilities.

MANUAL OF WATER SUPPLY PRACTICES—M20, Second Edition

Water Chlorination/Chloramination Practices and Principles

Disclaimer
The authors, contributors, editors, and publisher do not assume responsibility for the validity of the contents or any consequences of their use. In no event will AWWA be liable for direct, indirect, special, incidental, or consequential damages arising out of the use of information presented in this book. In particular, AWWA will not be responsible for any costs, including, but not limited to, those incurred as a result of lost revenue. In no event shall AWWA's liability exceed the amount paid for the purchase of this book.

Project Manager: Mary Kay Kozyra
Production Editor: Carol Stearns
Manuals Coordinator: Beth Behner

Library of Congress Cataloging-in-Publication Data

Water Chlorination/chloramination practices and principles
p. cm. -- (AWWA manual ; M20)
Rev. ed. of: Water chlorination principles and practices. 1973.
ISBN 1-58321-408-9
1. Water--Purification--Chlorination. 2. Water--Purification--Chloramination. I. American Water Works Association. II. Water chlorination principles and practices. III. Series

TD491.A49 no. M20 2006
[TD462]
628.1 s--dc22
[628.1'662] 2005057037

Printed in the United States of America
American Water Works Association
6666 West Quincy Avenue
Denver, CO 80235
ISBN 1-58321-408-9

Printed on recycled paper

Contents

Figures

This page intentionally blank.

Tables

Foreword

This publication is the second edition of the American Water Works Association (AWWA) Manual M20; the original manual was titled *Water Chlorination.* This manual provides information on chlorination and chloramination practices for the disinfection of drinking water. It does not cover the use of chlorine as an oxidant or as a disinfectant for pipeline and appurtenances.

The first edition was published in 1973 and was prepared to accompany an AWWA workshop on chlorination, which was part of a comprehensive operator's training program developed by an AWWA *ad hoc* committee and the AWWA Office of Education.

Major portions of the second edition have been adapted from two AWWA publications: *Water Treatment*, second edition, Principles and Practices of Water Supply Operations; and the *Chlorination / Chloramination Handbook* by G.F. Connell. In addition, significant contributions were obtained (with permission) from publications of the Chlorine Institute, the US Environmental Protection Agency, the Awwa Research Foundation, and the *Handbook of Chlorination and Alternative Disinfectants*, fourth edition, by Geo. Clifford White. AWWA thanks the authors and publishers of these resources for the use of their materials.

This page intentionally blank.

Acknowledgments

A working group was formed to develop this edition of AWWA Manual M20, *Water Chlorination / Chloramination Practices and Principles*. The working group, which consists of experts in various areas of chlorination, was authorized by the AWWA Disinfectants Standards Committee (213). It is due to their efforts that the first edition, published in 1973, was revised.

Manual Workgroup Members

W.B. Huebner, U.S. Filter/Wallace & Tiernan, Vineland, N.J.	(AWWA)
Kevin Kinback, U.S. Filter/Wallace & Tiernan, Vineland, N.J.	(AWWA)
Gary F. Trojak, The Chlorine Institute, Arlington, Va.	(AWWA)
Nicole K. Blute, McGuire Malcolm Pirnie, Inc., Santa Monica, Calif.	(AWWA)
Bill Lauer, AWWA, Denver, Colo.	(AWWA)

Wayne Huebner served as overall editor for the manual and chair of the workgroup. The workgroup gives special thanks to Michael McGuire, McGuire Malcolm Pirnie, Inc., for his review and input regarding chapter 6.

The second edition has been reviewed and approved by the AWWA Disinfectants Standards Committee (213). The chair of the committee, K. Blake Stark, thanks the contributing authors responsible for new material supplied to create an up-to-date chlorination practices manual. The committee members are listed below:

K. Blake Stark, Chair

G.F. Trojak, Vice Chair

General Interest Members

P.E. Cote, Cumberland, Maine	(AWWA)
D.J. Gates, Sabre Oxidation Technologies Inc., Citrus Heights, Calif.	(AWWA)
C.M. McLain,* Moorhead Public Service, Moorhead, Minn.	(AWWA)
S.J. Posavec,* Standards Group Liaison, AWWA, Denver, Colo.	(AWWA)
K.B. Stark, NSF International, Ann Arbor, Mich.	(AWWA)

*Liaison, nonvoting.

Producer Members

Dave Hildebrand,* Vulcan Chemicals, Wichita, Kan.	(AWWA)
W.B. Huebner, U.S. Filter/Wallace & Tiernan, Vineland, N.J.	(AWWA)
Thomas Nier, Bay Chemical & Supply Company, Odem, Texas	(AWWA)
G.F. Trojak, The Chlorine Institute, Arlington, Va.	(AWWA)

User Members

Ileana Alvarez,† John Preston Water Plant, Hialeah, Fla.	(AWWA)
B.S. Aptowicz, Philadelphia Water Department, Philadelphia, Pa.	(AWWA)
H.F. Hanson, Woodbridge, Va.	(AWWA)
A.A. Khan, Pepsi Cola North America, Plano, Texas	(AWWA)
R.C. Lorenz, Westerville Water Plant, Westerville, Ohio	(AWWA)
A.T. Segars, Miami-Dade Water and Sewer Department, Hialeah, Fla.	(AWWA)

*Deceased.
†Alternate.

Chapter 1

History of Chlorination and Chloramination

Chlorine was "discovered" in 1744 in an obscure laboratory in Sweden (AWWA 1973). Sixty-six years later in 1810, chlorine was identified as a chemical element and was named from the Greek word *chloros* (pale green), because of its characteristic color. However, it was not until several decades later that its use as a disinfectant was recognized.

Until the germ theory of disease was established around the mid-1880s, odors were thought to be responsible for transmitting diseases. As such, it was widely believed that the control of odors would limit the spread of infections. There is no question that the earliest applications of chlorine were aimed at controlling foul odors (Baylis 1935), even though, because of the unsophisticated techniques used, the chlorinous odor imparted to the treated water was sometimes just as, or more, objectionable than the odor it was meant to remove. It was not until well into the 1890s that chlorine and chlorine-containing products were evaluated and demonstrated to be effective disinfectants.

ORIGIN OF WATER DISINFECTION

Undoubtedly the disinfection of water has been practiced for millenniums, although probably with little or no understanding of the principles involved. Historical records indicate that the boiling of water had been recommended at least as early as 500 B.C. The earliest recorded use of chlorine directly for water disinfection was on an experimental basis, in connection with filtration studies in Louisville, Ky., in 1896. It was employed for a short period in 1897 in England, again on an experimental basis, to sterilize water distribution mains following a typhoid epidemic.

The first *continuous* use of chlorine for water disinfection was in 1902 in Belgium for the dual objective of aiding coagulation and making water biologically "safe." In North America, the first continuous, municipal application of chlorine to water was in 1908 to disinfect the 40-mgd (151-ML/d) Boonton Reservoir supply of the Jersey City, N.J., water utility. Contested in the courts, the practice was declared to represent a public health safeguard, and this action paved the way for its rapid extension to other public water supplies throughout North America and elsewhere.

RATIONALE OF WATER DISINFECTION

Waterborne diseases, such as typhoid, dysentery, and cholera, occurred with regularity in US water systems in the 1800s. The incidence rates were high in the early 1900s (e.g., more than 25,000 typhoid deaths in 1900) but decreased rapidly with the onset of chlorination (e.g., less than 20 deaths in 1960) (Laubusch 1964), as shown in Figure 1-1. The data provide similar results for other organisms.

With such a dramatic drop in illnesses and fatalities following the onset of chlorination, the need for feeding these chemicals was well justified. To date, chlorine has emerged as the disinfectant of choice primarily because of its effectiveness (Race 1918), efficiency, economy of operation, convenience, and the persistence of a chlorine residual. The unique properties of chlorine also make its use an acceptable technique for taste-and-odor control, algae and slime control, water-main disinfection, and many other purposes. However, chlorine has come under intense scrutiny because some of the by-products of the disinfection process may be carcinogenic. In spite of this, chlorine is expected to remain in wide use.

Water disinfection, as now ordinarily considered, involves specialized treatment for the destruction of harmful, and otherwise objectionable, "nuisance" organisms. Classically, disinfection processes have been employed to destroy or inactivate disease-producing (pathogenic) organisms—more particularly bacteria of intestinal origin. Such organisms can survive for weeks in the environment at temperatures near 70°F (21°C), or possibly even for months at lower temperatures. In addition to temperature influences, their survival in water depends on environmental, physiological, and morphological factors, such as pH, oxygen and nutrient supply, dilution, competition with other organisms, resistance to toxic influences, and ability to form spores. Whether these organisms actually will cause disease in humans upon ingestion depends on their virulence and concentration and also on the individual's vulnerability or susceptibility to them.

*Based on Vital Statistics of the United States 1937, 1938, 1943, 1944, 1949, 1960, 1967, 1976, 1987, 1992, Historical Statistics of the United States—Colonial Times to 1970 Part 1.

Source: HealthSentinel.com.

Figure 1-1 US typhoid mortality and disease rates

Water disinfection also encompasses the destruction of disease-producing organisms other than intestinal bacteria, but it does not necessarily imply the complete destruction of *all* living organisms (i.e., *sterilization*). Water disinfection processes seldom have been carried to the point of sterilization, which has been largely confined to the medical field. Among the other disease-producing organisms important in regard to water disinfection are a variety of viruses, intestinal protozoa, and some macroorganisms. In addition, many nuisance organisms of aesthetic or economic significance, both plant and animal, are sometimes vulnerable to disinfection and can be partially or completely controlled by suitable treatment.

EVOLUTION OF CHLORINATION MATERIALS

During the infancy of water chlorination, the only commercial sources of chlorine were dry chlorine-containing *compounds*, such as chlorinated lime (also called chloride of lime and bleaching powder) and sodium hypochlorite bleach solutions. The poor stability and variable effective chlorine content of the compounds then available caused many operating difficulties, and often inadequate disinfection dosages were used. Moreover, the comparatively crude equipment feeders yielded erratic results.

In 1909, liquid chlorine (the element in compressed form) became available commercially, and in the following year it was employed experimentally for water disinfection in Fort Myer, Va. The chlorine was added to the water by using a simple dry-gas feeder. In 1912, the first full-scale successful use of liquid chlorine was undertaken to control a recurring outbreak of typhoid in Niagara Falls, N.Y. In this case, a solution feeder was used. Two years later, improved equipment developed by C.F. Wallace and M.F. Tiernan to measure chlorine gas, dissolve it in water, and apply the solution was installed in Boonton, N.J., replacing the use of sodium hypochlorite bleach. These developments paved the way for the future extension of water disinfection techniques utilizing liquid chlorine.

Hypochlorite water chlorination gradually decreased in popularity, but it received renewed interest in 1928 with the commercial availability of high-test calcium hypochlorite, which was a more stable and active material than the various bleaching powders and solutions previously available. Today, there are three principal sources of chlorine for water disinfection and other chlorination treatment: (1) in elemental form, as a liquefied compressed gas (in commerce); (2) as "high-test calcium hypochlorite"; and (3) as chlorine bleach solutions (not to be confused with elemental liquid chlorine or with water solutions of chlorine gas).

EVOLUTION OF CHLORINATION CONTROL PRACTICES

Improvements in chlorine disinfection materials and in chemical feeders substantially contributed to the popularity and widespread adoption of water chlorination. Originally disinfection dosages were based largely on the application of fixed amounts of chlorine. Soon it became apparent that provisions were not being made for the effects of variations in water quality and the fluctuation in chlorine demand.

Gradually, the concept of varying chlorine dosage on the basis of residual chlorine was established, and iodometric methods for qualitative and quantitative assay of residual chlorine were developed. The use of orthotolidine as a qualitative indicator of residual chlorine was proposed in 1909, and later its use was extended quantitatively by the development of colorimetric standards. Between 1917 and 1919, chlorine disinfection was established on a scientific basis when the suitability and reliability of the orthotolidine test for even the smallest water supplies were demonstrated. Since then, a better understanding of water chlorination processes has brought many refinements in that test and in the development of other tests, such as the orthotolidine-arsenite colorimetric test to distinguish forms of residual chlorine, amperometric differential titration, and various other differential chemical tests for available chlorine. In recent years, the use of orthotolidine reagent has been discontinued in favor of the safer and more stable *N,N*-diethyl-*p*-phenylenediamine (DPD) method. DPD is commonly used both for color comparison and titrimetric chlorine testing.

DISCOVERY OF TRIHALOMETHANES

As chlorination processes gained further acceptance, new discoveries were limited to refinements in the existing methods and chemistry (Connell 1996). It wasn't until the 1970s that trace organic chlorination by-products were discovered. In Holland, Rook identified that the reaction of chlorine with organic materials dissolved in water produced a class of compounds called trihalomethanes (THMs) (Connell 1996). The organics, usually humic and fulvic acids that originate in decaying vegetative growths, can be found in agricultural runoffs, aquifers, and natural vegetation.

The THMs were identified as possible cancer-causing agents, and their discovery in drinking water raised concerns about chlorination. Research began on the nature of the reactions that produce THMs, the concentrations of THMs considered unacceptable in drinking water, and methods to reduce or prevent their formation. The great rush of scientific work produced sufficient data to lead regulatory bodies in Europe and North America to set acceptable THM concentration levels in finished water and to determine ways to minimize THM formation in existing plants.

HISTORICAL DEVELOPMENT OF CHLORAMINE

It is likely that chloramines formed accidentally in wastewater and in waters containing natural ammonia for some time before this reaction was characterized. In the early 1900s, the chlorine-ammonia combination received attention when it was found that the cost of chlorination might be reduced if ammonia was added (Race 1918). The practice of chloramine treatment was adopted in 1916 at the treatment plant in Ottawa, Ont. The first installation in the United States was in 1917 in Denver, Colo. Both locations used ammonia and hypochlorite and noted improvements in taste and reductions in after-growth in the distribution system.

Additional utilities adopted the use of chlorine-ammonia in the years that followed, as listed in Table 1-1, and more than 400 utilities were using the process by 1938. Breakpoint chlorination was discovered in 1939, and thereafter, the use of chloramines declined. Chloramine usage was further reduced during World War II when ammonia was in short supply.

Chloramines were then used sparingly until the 1970s when trace organic chlorination by-products were discovered (THMs). As discussed earlier in this chapter, THMs are produced from the reaction of chlorine with natural organic materials (humic and fulvic acids) and were also identified as possible carcinogens.

Subsequent to the discovery of THMs in drinking water, chloramination received renewed attention because it produces less THMs than *free* chlorine under most conditions. Haloacetic acids (HAAs) were identified as chlorination disinfection by-products and a group of five of these compounds were regulated in 1998. This further supported the use of alternate disinfectant such as chloramines.

A common disinfection strategy is to provide primary disinfection with chlorine or another strong disinfectant and then use chloramine as the residual disinfectant. This combination reduces chlorine by-product formation that would occur in the distribution system but still satisfies the need to ensure microbiologically safe drinking water. The widespread use of chloramines may be affected in the future, however, as new information is available on the by-products of chloramination.

Table 1-1 Utilities with long experience of chloramines use

City	Approximate Start of Chloramination
Denver, Colo.	1917
Portland, Ore.	1924
St. Louis, Mo.	1934
Boston, Mass.	1944
Indianapolis, Ind.	1954
Minneapolis, Minn.	1954
Dallas, Texas	1959
Kansas City, Mo.	1964
Milwaukee, Wis.	1964
Jefferson Parish, La.	1964
Philadelphia, Pa.	1969
Houston, Texas	1982
Miami, Fla.	1982
Orleans Parish, La.	1982
San Diego, Calif.	1982

Source: Trussell and Kreft 1984.

EVOLUTION OF CHLORAMINATION MATERIALS

The first observations of chloramine formation were in waters that contained ammonia. In addition, wastewater disinfection with chlorine often resulted in the formation of chloramines. Ammonia was initially added to water as aqua ammonia (liquid ammonium hydroxide) or ammonium salt solutions. Precipitation of calcium from water near the point of application prevented the widespread use of compressed ammonia gas. This issue was resolved and the use of anhydrous ammonia (gas) is now commonplace. The four forms of ammonia currently in common use are anhydrous ammonia (NH_3), aqua ammonia (NH_4OH), ammonium chloride (NH_4Cl), and ammonium sulfate ($[NH_4]_2SO_4$).

EVOLUTION OF CHLORAMINATION CONTROL PRACTICES

Chloramine residual testing was conducted originally using the same methods used for free chlorine. The total residual was determined by iodometric titration. It wasn't until the use of orthotolidine that speciation of the various forms of residual chlorine could be determined. This led to improvements in the dosage control of both chlorine and ammonia. Subsequently, the development of amperometric differential titration led to an even better understanding of the chemistry of the chlorine residual compounds. Today DPD is the most common test indicator for chlorine residual analysis, and it is possible to use DPD to differentiate some chloramine compounds.

Improvements in ammonia testing have also been important to control excess ammonia and to verify optimum chlorine:ammonia feed ratios. The Nessler method was used as early as 1930 to analyze wastewater for ammonia. This method used the formation of iodine and its characteristic color to determine the concentration of ammonia. The phenate method provides an alternative when some interferences are present. The ammonia-selective electrode method, developed in the 1970s, has made the routine testing of ammonia in water possible.

DISINFECTION REGULATIONS IN THE UNITED STATES

In 1974, the United States passed the Safe Drinking Water Act (SDWA). Amendments were added in 1986 and again in subsequent years that set further criteria for acceptable levels of THMs (Connell 1996). The SDWA also encouraged alternate treatment methods to achieve the now twin goals of disinfection and minimized THMs.

The Disinfectants and Disinfection By-Products Rule (USEPA 2001) established a maximum contaminant level (MCL) for THMs of 80 µg/L. The same rule set the MCL for HAAs at 60 µg/L. Maximum residual disinfectant levels were also set at 4.0 mg/L (ppm) for both chlorine and chloramine.

These regulations have caused utilities to carefully manage the use of chlorine. For example, the use of chlorine at the raw water intake is being reconsidered in favor of moving the point of addition closer to the clearwell. Intermediate points in the treatment plant would be considered on an as-needed basis. The addition of chlorine in conjunction with ammonia for use as the residual disinfectant in the distribution system is gaining popularity.

INTERNATIONAL DISINFECTION REGULATIONS

The World Health Organization (WHO) is a specialized agency of the United Nations. It publishes guidelines for drinking water quality that are used as a basis for the development of national drinking water standards. The 1998 guidelines include limits for disinfectants and disinfection by-products.

The European Union (EU) issued a directive (98/83/EC) that established minimum standards for water intended for human consumption. (More strict standards within the member states are allowed.) The directive includes disinfectants and disinfection by-product limits similar to the WHO guideline values.

The Netherlands has water systems operating without a residual in the distribution systems, indicating its great concern about the formation of THMs and other chlorinated by-products. The Netherlands' practices are not generally shared by the other European members. However, most of the practices of the EU states in the treatment of surface waters involve pretreatment with either ozone or chlorine dioxide, followed by chlorination of the finished waters.

Many countries have established their own drinking water quality standards. Most of these have adopted USEPA, EU, WHO, or a combination for their standards, yet the enforcement of standards varies widely from entirely voluntary to strict governmental oversight.

THE FUTURE

Chlorination practices will continue to be under scrutiny as more is learned about the effects of the disinfection process and the resulting disinfection by-products. Operation of surface water treatment plants and the quality of the water they produce will continue to be followed closely. Treated, potable water will continue to be examined in ever-increasing detail. New processes that will remove organic precursors will be discovered and used successfully.

Disinfection of drinking water is a vital component of the treatment and delivery system, and the protection of public health is the primary function of drinking water purveyors. It is a great challenge to provide microbiologically safe water and at the same time control disinfectant by-product production. Currently, scientific information is available to achieve these competing objectives. Further advancements, from scientific research, will undoubtedly lead the use of chlorine and chloramines in new directions in the future.

REFERENCES

Baylis, J.R. 1935. *Elimination of Taste and Odor in Water.* New York: McGraw-Hill, Inc.

Connell, G.F. 1996. *The Chlorination/Chloramination Handbook.* Denver, Colo.: American Water Works Association.

Laubusch, E.J. 1964. *Chlorination and Other Disinfection Processes.* Washington, D.C.: The Chlorine Institute.

Race, J. 1918. Chlorination and Chloramines. *Jour. AWWA*, 3:63.

Trussell, R., and P. Kreft. 1984. Engineering Considerations of Chloramine Application. AWWA, Seminar Proc.

US Environmental Protection Agency. 2001. *Stage 1 Disinfectants and Disinfection Byproducts Rule: A Quick Reference Guide.* EPA 816-F-01-010. Washington, D.C.: USEPA Office of Water.

ADDITIONAL SOURCES OF INFORMATION

APHA, AWWA, and WEF (American Public Health Association, American Water Works Association, and Water Environment Federation). 1999. *Standard Methods for the Examination of Water and Wastewater,* 20th ed. Washington, D.C.: APHA.

AWWA, 1999. *Water Quality and Treatment*, 5th ed. R.D. Letterman, ed. New York: McGraw-Hill.

This page intentionally blank.

Chapter 2

Properties of Chlorination Chemicals

CHLORINE GAS

In the gaseous state, chlorine is greenish-yellow in color and has a pungent odor. The chemical symbol for chlorine is Cl and its atomic weight is 35.46. Chlorine, however, exists only as a two-atom molecule that is identified by the symbol Cl_2 and has a molecular weight of 70.92. It is neither explosive nor flammable, although as a strong oxidant, it supports combustion and reacts violently with many substances.

Pure chlorine does not exist naturally because of its reactivity. It is commercially produced by the electrolysis of sodium chloride brine. The chemical reaction is

$$\underset{\text{sodium chloride}}{2NaCl} + 2H_2O + \text{electric current} \rightarrow \underset{\text{sodium hydroxide}}{2NaOH} + \underset{\text{chlorine}}{Cl_2} + \underset{\text{hydrogen}}{H_2} \tag{2-1}$$

Chlorine is stored and shipped as a liquefied gas under pressure, usually in 100- to 150-lb (45- to 68-kg) cylinders, ton containers, and tank cars. The compressed liquid is amber in color and approximately 1.5 times as heavy as water. If liquid chlorine is unconfined, it rapidly vaporizes to a gas. (One liquid volume yields approximately 450 volumes of gas.) The conversion of liquid to gas and subsequent expansion requires heat and can be self limiting by heat transfer to the liquid. As such, heat transfer through container walls limits gas withdrawal rates.

Because chlorine in a container may exist in gas, liquid, or both forms, any consideration of liquid chlorine includes that of chlorine gas. Chlorine gas is approximately 2.5 times as heavy as air. Therefore, released gas will initially settle to the lowest point in the container until mixed with the surrounding air. Figure 2-1 illustrates the temperature–pressure characteristics of chlorine. Note that when chlorine temperature increases, pressure also increases.

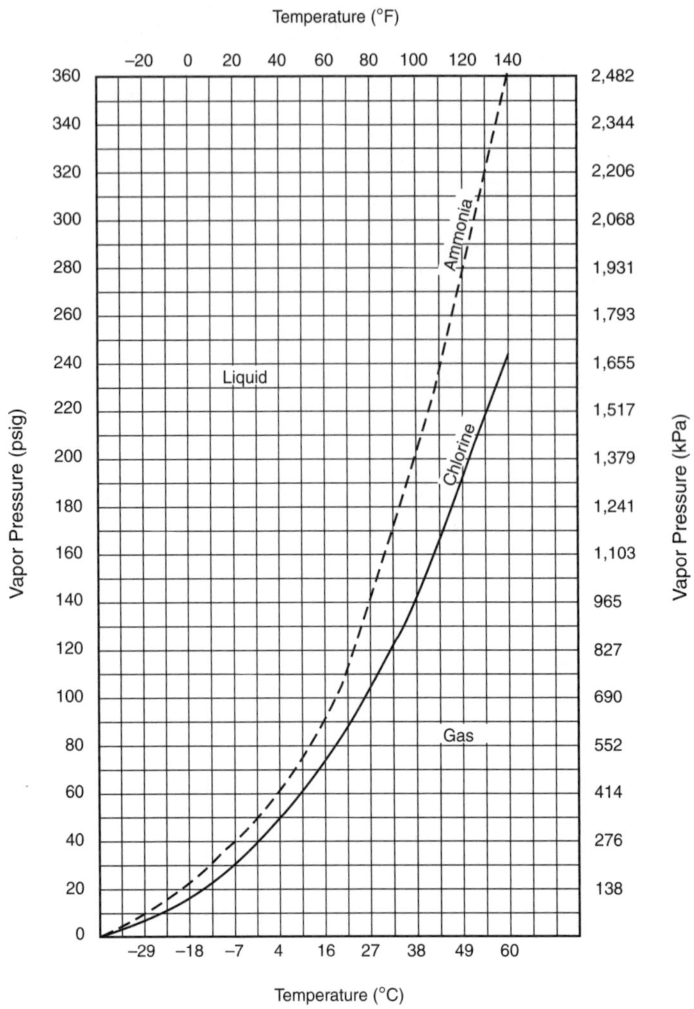

Source: Connell, 1996.

Figure 2-1 Vapor pressure of liquid chlorine

Figure 2-2 illustrates the volume–temperature characteristics of chlorine in a container loaded to the maximum authorized by the US Department of Transportation (USDOT). Note that a shipping container filled to its authorized maximum becomes completely full of liquid at approximately 154°F (68°C). Temperatures beyond that point may produce pressure that could result in hydrostatic rupture of the container. As such, safety devices are an integral part of chlorine containers. Fusible plugs are designed to melt at temperatures between 158 and 170°F (70 and 77°C), relieving excessive pressure accompanying temperature increases.

Chlorine is only slightly soluble in water; its maximum solubility is approximately 1 percent (10,000 ppm [mg/L]) at 49.3°F (9.6°C). Because the vapor pressure of chlorine increases with rising temperature, its solubility also decreases. The solubility drops to 4,000 ppm at 100°F (38°C) and to 0 ppm at 212°F (100°C). At temperatures below 49.3°F (9.6°C), chlorine can combine with water to form chlorine hydrate (sometimes called chlorine "ice"), which is a crystalline substance that may cause operational problems with direct feed chlorination. The maximum solubility for a given temperature should only be used for reference purposes. The actual useful solubility with minimal vapor pressure is considerably less: 3,500 mg/L is typically used as the maximum practical limit.

Table 2-1 summarizes some of the important physical properties of chlorine.

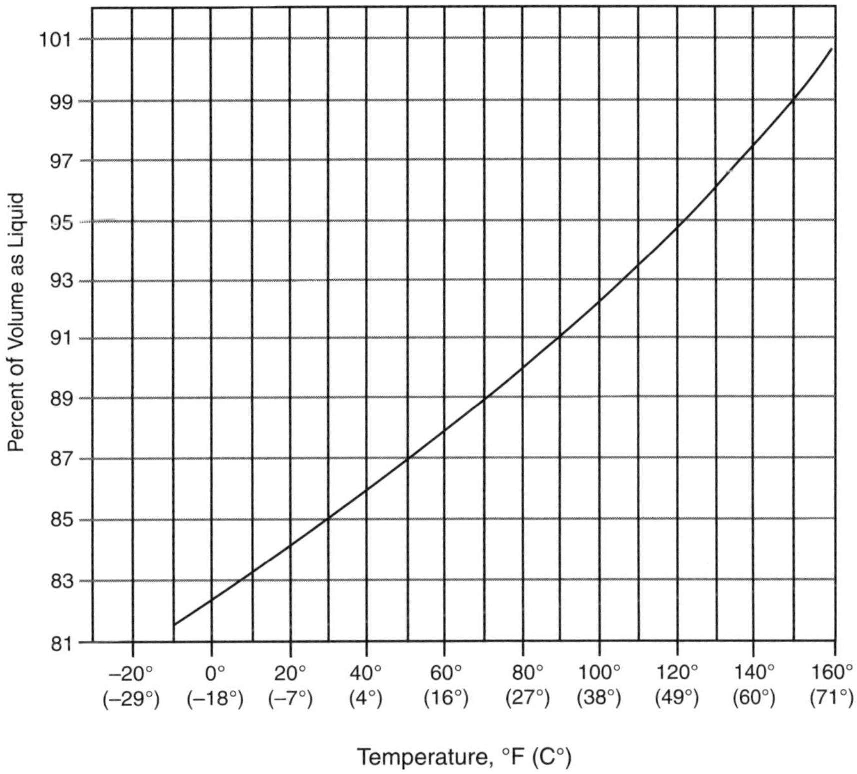

Adapted from the Chlorine Institute Chlorine Manual.

Figure 2-2 Volume–temperature relation of liquid chlorine in a container loaded to its authorized limit

Table 2-1 Physical properties of chlorine

Boiling point (liquefying point) at 1 atm*	–29.15°F (–33.97°C)
Melting point (freezing point) at 1 atm	–149.76°F (–100.98°C)
Liquid density at 60°F (16°C)	88.8 lb/cu ft (1,422 kg/m^3)
Gas density at 34°F (1.1°C)	0.2006 lb/cu ft (3.213 kg/m^3)
Specific gravity (liquid) at 32°F (0°C)	1.468 (water = 1)
Specific gravity (gas) at 32°F (0°C)	2.485 (air = 1)
Water solubility at 70°F (21.1°C)	0.7 percent
Vapor pressures:	
at 32°F (0°C)	53.51 psi (368.9 kPa)
at 77°F (25°C)	112.95 psi (778.8 kPa)
at 129°F (43.9°C)	191.01 psi (1,316.8 kPa)

*1 atmosphere = 14.696 psi (101.325 kPa).

Chemical Properties

Although it is neither explosive nor flammable, chlorine is a strong oxidant and capable of supporting combustion of certain substances. It should be handled and stored away from other compressed gases (e.g., ammonia) and flammable materials.

When mixed with moisture or water, chlorine will hydrolyze to form a corrosive mixture of acids and strong oxidants shown as

$$\underset{\text{chlorine}}{Cl_2} + H_2O \leftrightarrow \underset{\text{hypochlorous acid}}{HOCl} + \underset{\text{hydrochloric acid}}{HCl} \qquad (2\text{-}2)$$

The chemical reactivity of chlorine enables it to combine with many other chemicals. It reacts with most organic materials, sometimes with explosive results. Dry chlorine is not corrosive to metals such as carbon steel, nickel, lead, or copper, but it can react violently with other metals such as aluminum, tin, gold, and titanium. Stainless steels are subject to chloride stress corrosion and should not be used by water utilities for chlorine service. Moist chlorine is corrosive to most metals with the exception of gold, silver, platinum, titanium, and certain specialized alloys.

Some plastics, such as polytetrafluoroethylene (PTFE) and polyvinylidene fluoride (PVDF), can be used with gaseous chlorine whether wet or dry. Dry liquid chlorine or gas under pressure will react with polyvinyl chloride (PVC), causing it to soften and reduce its structural integrity. PTFE and PVDF may be used in liquid chlorine service. Gaseous chlorine under vacuum, both wet or dry, is compatible with PVC, chlorinated polyvinyl chloride (CPVC), or acrylonitrile butadiene styrene (ABS).

SODIUM HYPOCHLORITE

Sodium hypochlorite (NaOCl), often referred to as liquid bleach, is available only as a solution that contains 5–20 percent available chlorine. These solutions are clear to light yellow in color, are alkaline corrosive, and have a strong chlorinous odor. (The vapor above solutions contains a mixture of hypochlorous acid and chlorine monoxide.) The percent concentration typically used with hypochlorite is in terms of "trade percent." This is different from a "weight percent," which takes into consideration the solution's specific gravity. Trade percent is expressed as

$$\text{trade percent} = \frac{\text{chlorine (g/L)}}{1{,}000} \times 100 \qquad (2\text{-}3)$$

A 15 percent solution contains 1.5 lb/gal (180 g/L) of chlorine; a 10 percent solution contains 1 lb/gal (100 g/L), and so on.

Sodium hypochlorite is produced by reacting chlorine and sodium hydroxide as follows:

$$\underset{\text{chlorine}}{Cl_2} + \underset{\text{sodium hydroxide}}{NaOH} \rightarrow \underset{\text{sodium hypochlorite}}{NaOCl} + \underset{\text{sodium chloride}}{NaCl} + \text{heat} \qquad (2\text{-}4)$$

As noted earlier in this chapter, heat is generated during this reaction, which must be controlled to minimize the formation of chlorate at the onset and maximize stability during storage.

Unlike elemental chlorine, sodium hypochlorite is subject to decomposition. Table 2-2 lists the primary factors that affect stability.

Table 2-2 Primary factors affecting sodium hypochlorite stability

Factor	Relationship	Decomposition Products
Concentration	Stability increases inversely with concentration	Chlorate
Temperature	Stability increases inversely with temperature	Chlorate
Metallic impurities	Minimize Cu, Ni, Co concentrations	O_2
pH	A pH between 11.5 and 13 is best	Chlorate
UV light	Minimize light exposure	Chlorate/O_2

Based on the influencing factors listed in Table 2-2, the rate of decomposition of liquid hypochlorite has been studied (Gordon et al. 1995; Bommaraju 1994), and several equations have been developed to help predict the decomposition rate. (For additional details see the *Sodium Hypochlorite Manual* [Chlorine Institute 1995].) While helpful, these equations rely on a number of assumptions as well as accurate and consistent information, all of which may not be available. The best method to use in dealing with decomposition is to minimize the negative factors outlined in Table 2-2 and monitor the solution concentration. For a crude estimate, under moderate conditions, a 15 percent hypochlorite solution can conservatively be expected to degrade by approximately 0.5 percent per day. For example, 15 percent sodium hypochlorite at 77°F (25°C) has a 100-day half-life (i.e., it will degrade to half of its strength in 100 days). For comparison, at the same temperature, the half-life of 10 percent sodium hypochlorite is 220 days, and 5 percent is 790 days.

The three implications to sodium hypochlorite decomposition include

- the slow degradation of feed concentration,
- possible health consequences of high chlorate levels, and
- the evolution of oxygen that may complicate the feed.

Because sodium hypochlorite solution concentration degrades with time, feed volumes must be increased to compensate. This is easily accomplished by adjusting the volumes manually (once a week is adequate under most conditions), or by providing residual feedback control as outlined in chapter 4. Concentration of new shipments of sodium hypochlorite should be checked daily for the first few days until stability can be characterized.

Chlorate is a by-product of decomposition, formed in the manufacturing process and during storage, as illustrated here:

$$\underset{\text{hypochlorite ion}}{3OCl^-} \rightarrow \underset{\text{chlorate}}{ClO_3^-} + \underset{\text{chloride}}{2Cl^-} \quad (2\text{-}5)$$

The human health risk of chlorate has not been clearly defined in any definitive studies. For example, the 2002 edition of the *Drinking Water Standards and Health Advisories* (USEPA 2002) does not contain a listing for chlorate. However, high levels of chlorate have been associated with hemolytic anemia. Health Canada has proposed a guideline for chlorate in drinking water of 1.0 mg/L and the World Health Organization has a provisional guideline of 0.7 mg/L.

A second major degradation path, primarily catalyzed by transition metals such as Ni^{2+}, Cu^{2+}, and Co^{2+}, produces oxygen as follows:

$$\underset{\text{hypochlorite}}{OCl^-} + \underset{\text{hypochlorite}}{OCl^-} \rightarrow \underset{\text{oxygen}}{O_2} + \underset{\text{chloride}}{2Cl^-} \quad (2\text{-}6)$$

While oxygen is not toxic, this degradation pathway can dramatically reduce the solution half-life and gas-bind chemical feed pumps resulting in cavitations and irregular feed rates. Metal concentrations in the sub-ppm levels can be a problem. Contamination can result from the manufacturing process, storage, transportation, feeding, or via dilution water. Particulates can be a major source of metal contamination, and filtration during the storage tank fill process can be an effective preventative measure. All metal (with the exception of titanium and tantalum) contact must be eliminated from storage vessels, piping, valves, and feed equipment.

Steps to Reduce Hypochlorite Degradation

Regardless of any other factors, sodium hypochlorite solutions will degrade with time. Any steps to minimize the time between manufacture, delivery, and use will maximize product strength and reduce chlorate formation. These steps will, however, involve trade-offs in product cost and operation time between deliveries.

The products of decomposition of most concern are chlorate and oxygen. There is no specific data showing that the expected levels of chlorate from hypochlorite would be hazardous to human health. However, it is prudent to minimize the formation of unintended by-products that may be of future concern. Oxygen, on the other hand, while not a toxicity issue, can complicate feeding.

A small reduction in hypochlorite solution concentration will greatly increase solution stability, as outlined in Table 2-3. However, care must be exercised to dilute hypochlorite solutions only with good quality waters. Using water that has been filtered and passed through a softener (cation exchange resin) to remove any solubilized metals is a good precaution. The pH must also be between 11.5 and 13.

Decreasing the temperature of hypochlorite solutions also has substantial benefit. As shown in Table 2-3, a 59°F (15°C) decrease in temperature increased stability over six times. While solution temperature is primarily a result of storage conditions, and options to reduce temperature may be limited, any reduction in temperature will have a measurable positive effect on solution stability. Such options may include

- specifying a maximum temperature of delivered hypochlorite,
- avoiding daylight heating by covering storage containers and feed piping, and
- considering practical heat exchangers and cooled storage areas.

Both oxygen and chlorate formation are catalyzed by ultraviolet (UV) light (sunlight). At the very least, the storage facility should be covered, preferably with reflective material, and storage containers made of or covered with opaque or UV blocking material.

Table 2-3 Half-life values of liquid bleach: varying temperature, pH, and concentration

Temperature		9.5 wt%	4.25 wt%	2.13 wt%
(°F)	(°C)	pH 13.0	pH 12.7	pH 12.4
95	35	66 days	300 days	2 years
68	20	1.5 years	6.5 years	17 years

Utilities should also establish minimum specifications for sodium hypochlorite deliveries to help in minimizing degradation products and rates. A hypochlorite specification may include the following requirements:

- a pH greater than 12,
- Ni^{2+} and Cu^{2+} concentrations less than 0.1 mg/L,
- a maximum chlorate concentration,
- a concentration greater than 12.5 percent (consider a lower concentration if cost per pound is comparable and feed and storage capacity can accommodate), and
- a solution temperature prior to delivery that is below 77°F (25°C).

The enforcement of a specification will greatly assist solution stability and predictability of degradation rate. For more detailed specification information, readers are directed to the AWWA Standard for Hypochlorites (B300, latest edition).

AMMONIA GAS

Ammonia is a gas at atmospheric pressure and ambient temperatures. It is not an element but rather a compound composed of the elements nitrogen and hydrogen. The chemical formula for ammonia is NH_3 and contains three atoms of hydrogen and one atom of nitrogen. Its molecular weight is 17. Small quantities of ammonia exist naturally. For commercial purposes, ammonia can be synthesized by a number of different processes that use hydrogen and nitrogen.

Ammonia has a pungent odor and, unlike chlorine, is colorless in both gaseous and liquid states. Ammonia liquefies at approximately the same temperature as chlorine but has a higher vapor pressure than chlorine at corresponding temperatures. The vapor pressure–temperature curve for ammonia is shown in Figure 2-1. To illustrate the difference in vapor pressure between chlorine and ammonia with changing temperatures, the dotted line in the figure represents the ammonia vapor pressure curve. The vapor pressure of ammonia is approximately 20 percent greater than chlorine at ambient temperatures and increases at a more rapid pace than chlorine at higher temperatures.

Unlike chlorine, reliquefaction of ammonia gas is not considered to be an operational problem. The boiling point, or vaporization point, of ammonia at atmospheric pressure is –28°F (–33.3°C).

Liquid ammonia is lighter than water with a density of 42.57 lb/ft^3 (681.9 kg/m^3) at –18°F (–28°C). Ammonia gas is lighter than air with a density of 0.555 lb/ft^3 (0.8899 kg/m^3) at –18°F (–28°C). Ammonia is approximately 36 times more soluble in water than chlorine: 250 lb/100 gal (299 g/L) or 34 percent by weight at 60°F (20°C).

Ammonia is classified by the USDOT as a corrosive gas. Chemically, ammonia is a relatively stable chemical compound and reactive only under specific conditions or with certain chemicals. Most common metals are unaffected by dry ammonia. In the presence of moisture, or when dissolved in water, ammonia will attack copper, zinc, and alloys containing these metals (e.g., brass and bronze). Ammonia will react with organics and inorganics to form ammonium salts. Reactions with chlorine can produce dangerous and/or explosive compounds, such as nitrogen trichloride. The physical properties of ammonia are listed in Table 2-4.

Ammonia is considered a respiratory irritant, although it does not have a cumulative effect. Ammonia's pungent odor provides an early warning to alert the individual to its presence. Although individual physiological responses may vary, the least perceptible odor is considered to be 5 ppm (by volume). The current (1993) OSHA-PEL level is 35 ppm, while the American Council of Government Industrial Hygienists (ACGIH) threshold limit value and short-term exposure limit (STEL) value are 25 ppm and 35 ppm, respectively.

Table 2-4 Physical properties of ammonia

Molecular Weight	17.03
Boiling point	–28.2°F (–33.4°C) at 1 atm
Freezing point	–107.9°F (–77.7°C)
Density (gas)	0.0555 lb/ft^3 (0.8899 kg/m^3) at –27.7°F (–33.2°C)
Density (liquid)	42.57 lb/ ft^3 (681.9 kg/m^3) at –27.7°F (–33.2°C)
Flammability	Only within a range of 16–25 percent

Source: Compressed Gas Association 1984.

Table 2-5 Properties of aqueous (aqua) ammonia, 30 percent by weight

Specific gravity at 60°F (15°C)	0.8957
Density at 60°F (15°C)	26.31°Bé (Baumé) 55.7 lb/ft^3 (0.893 kg/L)
Boiling point at 1 atm (101.325 kPa)	82°F (27.8°C)

AMMONIA SOLUTIONS

Ammonia is available in forms other than as a compressed gas in cylinders. The most commonly used form is a solution of ammonia dissolved in water, usually referred to as aqueous or aqua ammonia. Other forms of ammonia are salts of ammonia solutions. Of these, the most frequently used is ammonium sulfate. The reaction of ammonia with water is shown here:

$$\underset{\text{ammonia}}{NH_3} + H_2O \leftrightarrow \underset{\text{ammonium hydroxide}}{NH_4OH} \tag{2-7}$$

The product of this reaction is ammonium hydroxide, which is also produced from solutions of ammonia salts, as shown here:

$$\underset{\text{ammonium sulfate}}{(NH_4)_2SO_4} + 2H_2O \rightarrow \underset{\text{ammonium hydroxide}}{2NH_4OH} + \underset{\text{sulfuric acid}}{H_2SO_4} \tag{2-8}$$

Equation 2-7 is reversible and will release ammonia as a gas. In addition, solutions that produce ammonium hydroxide will also provide ammonia. Because ammonia is available from these solutions, ammonium salt solutions are used as a source of ammonia just as sodium hypochlorite is used as a source of chlorine. All ammonium salt or aqueous ammonia solutions have alkaline pH values. (The exact pH is a function of the solution concentration and temperature.) Because some ammonia gas evolves, a vapor pressure exists in aqueous ammonia and ammonium salt solutions, and all storage facilities must be suitably vented. Table 2-5 lists the properties of aqueous ammonia.

Both ammonium salt solutions and aqueous ammonia exhibit the same characteristics that would be expected from ammonia. The solutions should be treated as a respiratory irritant and handled with care. Irritation and redness may develop if the solutions come into contact with the skin.

For additional information on the properties and handling of chlorine, ammonia, and their solutions, see chapter 6. Also ask the chemical supplier to provide the latest available material safety data sheet (MSDS) for the appropriate chemical.

REFERENCES

Bommaraju, T.V. 1994. Sodium Hypochlorite: Its Chemistry and Solubility. Chlorine Institute Packaging Seminar, Minneapolis, Minn.

Chlorine Institute. 1995. *Sodium Hypochlorite Manual*. Arlington, Va.: The Chlorine Institute.

Compressed Gas Association. 1984. *Anhydrous Ammonia, Pamphlet G-2*. Alexandria, Va.: Compressed Gas Association.

Connell, G.F. 1996. *The Chlorination/Chloramination Handbook*. Denver, Colo.: American Water Works Association.

Gordon, G., L. Adam, and B. Bubnis. 1995. *Minimizing Chlorate Ion Formation in Drinking Water When Hypochlorite Ion Is the Chlorinating Agent*. Denver, Colo.: American Water Works Association.

USEPA (US Environmental Protection Agency). 2002. *Drinking Water Standards and Health Advisories*. EPA 822-R-02-038. Washington, D.C.: USEPA, Office of Water.

ADDITIONAL SOURCES OF INFORMATION

Chlorine Institute, Inc. 1996. 1997. *Chlorine Manual*. Pamphlet 1, 6th ed. Arlington, Va.: The Chlorine Institute, Inc.

———. *Molecular Chlorine: Health and Environmental Effects*. Pamphlet 90. Arlington, Va.: The Chlorine Institute, Inc.

———. 2000. *Sodium Hypochlorite Manual*. Pamphlet 96, 2nd ed. Arlington, Va.: The Chlorine Institute, Inc.

Compressed Gas Association. 1984. 1995. Anhydrous Ammonia. Pamphlet G-2, 8th ed. Chantilly, Va.: Compressed Gas Association.

LaRoche Industries. 1989. *Ammonia Technical Data Manual*. Duluth, Ga.: LaRoche Industries.

Storage and Handling of Anhydrous Ammonia, Technical Data, February 1989. National Ammonia Company.

This page intentionally blank.

Chapter 3

Chlorination Water Chemistry and Disinfection Mechanisms

CHLORINATION CHEMISTRY

When chlorine gas (Cl_2) is added to water, a mixture of hypochlorous acid (HOCl) and hydrochloric acid (HCl) is formed, as shown here:

$$\underset{\text{chlorine}}{Cl_2} + H_2O \rightarrow \underset{\text{hypochlorous acid}}{HOCl} + \underset{\text{hydrochloric acid}}{HCl} \tag{3-1}$$

In dilute solutions and at pH levels above 4, the reaction illustrated in Eq 3-1 is displaced to the right. As the pH drops below 4, the percentage of available chlorine in the form of chlorine gas increases along with the vapor pressure. For every milligram per liter (part per million) of chlorine added, 0.7–1.4 mg/L of alkalinity is consumed. At typical chlorine dose rates (1–5 mg/L), the pH of the receiving water will be reduced but buffered by the water's natural alkalinity.

Once formed the HOCl instantaneously establishes equilibrium as follows:

$$\underset{\text{hypochlorous acid}}{HOCl} \leftrightarrow H^+ + \underset{\text{hypochlorite ion}}{OCl^-} \tag{3-2}$$

As the pH rises above 7.5 (60°F [20°C]), an increasing percentage of free chlorine is in the form of hypochlorite ion. At a pH below 7.5, free chlorine is in the form of HOCl, as illustrated in Figure 3-1. As shown, temperature will also influence the equilibrium; the effect is particularly significant at a pH between 7 and 8. This reaction is independent of concentration, is completely reversible, and responds instantaneously to pH changes.

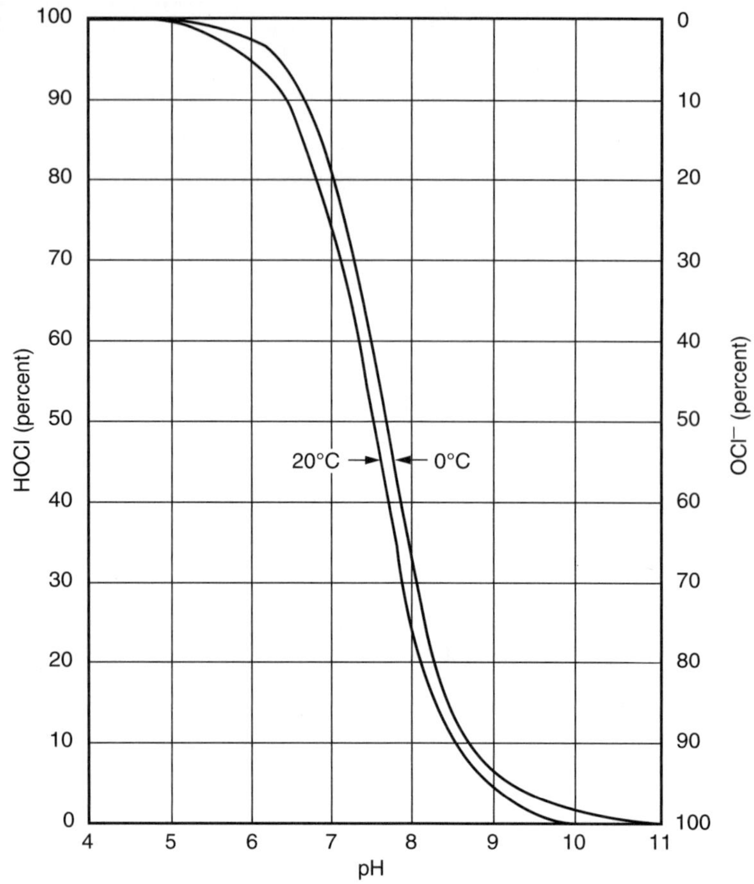

Source: Connell, 1996.

Figure 3-1 Hypochlorous acid/hypochlorite distribution versus pH

The combined concentration of HOCl and OCl^- is known as *free chlorine residual* (FCR). These two species, however, react quite differently. HOCl is a much stronger disinfectant, stronger oxidant, and more reactive than OCl^-. HOCl will disinfect 100 times faster and oxidize compounds that OCl^- will not, but be consumed at a much higher rate. This creates a dynamic relationship in which HOCl is consumed and the remaining FCR re-equilibrates according to the relationship illustrated in Figure 3-1. In effect, OCl^- acts as a disinfectant buffer, adding HOCl as it is being consumed.

The speed of chlorine disinfection and durability of the residual is therefore strongly affected by the pH of the water being treated. In addition to being more reactive, HOCl is neutral and can more easily penetrate negatively charged bacterial surfaces and suspended particles protecting pathogens.

Chlorine may also be added as sodium hypochlorite or calcium hypochlorite. While there are obvious differences in feeding characteristics, and hypochlorite will add alkalinity as opposed to chlorine gas consuming alkalinity, once added to the receiving water the resulting residuals are indistinguishable from those added by chlorine gas as shown here:

$$\underset{\text{sodium hypochlorite}}{NaOCl} + H_2O \rightarrow \underset{\text{hypochlorous acid}}{HOCl} + \underset{\text{sodium hydroxide}}{NaOH} \quad (3\text{-}3)$$

$$Ca(OCl)_2 + 2H_2O \rightarrow 2HOCl + Ca(OH)_2 \quad (3\text{-}4)$$

calcium hypochlorite; hypochlorous acid; calcium hydroxide

The reaction products of both chemicals produce HOCl, which will be distributed according to the HOCl and OCl^- pH relationship.

Chlorine Reactions With Other Compounds

The reactions of chlorine with other compounds are important to understand because these reactions generally exhibit a chlorine demand. This may directly affect the disinfection capability of the chlorine added to the water.

Reactions with ammonia. The reaction between chlorine and ammonia is of great significance in water chlorination processes, especially disinfection. When chlorine is added to water containing natural or added ammonia, the ammonia reacts with HOCl to form various chloramines that, like HOCl, retain the oxidizing capacity of chlorine (+1 oxidation state) but in a weaker configuration. The reactions between chlorine and ammonia may be represented as follows:

$$HOCl + NH_3 \rightarrow NH_2Cl + H_2O \quad (3\text{-}5)$$

hypochlorous acid; ammonia; monochloramine

$$HOCl + NH_2Cl \rightarrow NHCl_2 + H_2O \quad (3\text{-}6)$$

hypochlorous acid; monochloramine; dichloramine

$$HOCl + NHCl_2 \rightarrow NCl_2 + H_2O \quad (3\text{-}7)$$

hypochlorous acid; dichloramine; nitrogen trichloride

The distribution of the reaction products is governed by the rates of formation of monochloramine (NH_2Cl) and dichloramine ($NHCl_2$), which are dependent upon pH, temperature, time, and initial chlorine to ammonia ($Cl:NH_3$) ratio. In general, low $Cl:NH_3$ ratios and high pH levels favor monochloramine. Monochloramine is characterized as a weak disinfectant and oxidant, frequently used as a durable residual that will form lower levels of disinfection by-products (DBPs) and will produce the least detectable chloronous taste and odor (T&O) of any of the chlorine residuals. Dichloramine has similar disinfection and oxidation characteristics to monochloramine but is characterized by a more intense T&O. Nitrogen trichloride (NCl_3) is practically insoluble in water, has a very intense T&O, and is known for its eye irritation/tearing effects at very low concentrations.

Chlorine also reacts with organic nitrogenous material, such as proteins and amino acids, to form organic chloramines. These compounds have little disinfection and oxidation value and are resistant to further oxidation. These residuals are sometimes referred to as "nuisance residuals" because they are virtually indistinguishable from inorganic chloramines with standardized analytical procedures.

The breakpoint reaction. The breakpoint is described as the point at which chlorine demand has been satisfied, combined chlorine compounds have been destroyed, and as additional chlorine is added, a free chlorine residual is produced. The breakpoint reaction is important in understanding the formation of chloramines (when ammonia is present or is added) because the formation of chloramines is a primary element of this reaction. In some cases, organic compounds present in the water may also combine with chlorine to form chloro-organic compounds that may exhibit the characteristics of a combined chlorine residual.

If water contains ammonia (either naturally or added), and a free chlorine residual is desired, the ammonia can be removed by oxidation with chlorine to produce primarily nitrogen gas and some nitrate by the breakpoint reaction. While there are a number of reaction steps that take place in this process, two of the predominant reaction sequences are summarized as follows:

$$\underset{\text{ammonia}}{2NH_4^+} + \underset{\text{hypochlorous acid}}{3HOCl} \rightarrow \underset{\text{nitrogen}}{N_2} + 5H^+ + \underset{\text{chloride}}{3Cl^-} + 3H_2O \quad (3\text{-}8)$$

$$\underset{\text{ammonia}}{NH_4^+} + \underset{\text{hypochlorous acid}}{4HOCl} \rightarrow \underset{\text{nitrate}}{NO_3^-} + H_2O + \underset{\text{hydrogen ion}}{6H^+} + \underset{\text{chloride}}{4Cl^-} \quad (3\text{-}9)$$

Ideally, the breakpoint reaction should take place at a pH between 7.0 and 7.5, but it will also continue at a slower rate down to a pH of 6.5 and as high as 8.5. Adequate chlorine must also be added to complete the reaction (empirically found to be approximately 8.5 mg/L [ppm] chlorine per mg/L [ppm] NH_3-N). Additional chlorine may also be required to satisfy a number of side reactions, including the creation of minor products such as nitrogen trichloride, and other chlorine demands (caused by organic and other compounds). The only other reaction requirement is time. The breakpoint reaction is not instantaneous and may take 30 min or longer to run to completion. If free chlorine disinfection or oxidation is needed, adequate reaction time must be provided.

The ammonia breakpoint reaction is a series of steps in which monochloramine is formed first and then converted by HOCl to dichloramine, which is followed by a dichloramine decomposition reaction resulting in primarily nitrogen gas and nitrate. Organic compounds may combine with chlorine to form chloro-organic compounds that may also be destroyed as more chlorine is added.

The breakpoint curve illustrates the breakpoint reaction and is shown in Figure 3-2. This curve plots the chlorine dosage in milligrams per liter along the horizontal axis and the total (free and combined) chlorine residual in milligrams per liter along the vertical axis. The line plotted at a 45° angle is called the zero demand line and represents the ideal situation in which all chlorine added to the water is measured as a free chlorine residual in the water. In zero demand water, there is no loss of chlorine addition due to chlorine demand.

The presence of reducing agents and other inorganic chlorine demand-causing compounds will consume any initial chlorine dosage, which results in zero chlorine residual until this demand is satisfied. This section of the zero demand curve is identified as zone 1 in Figure 3-2. Once this initial demand is met in the presence of ammonia (and organic compounds), any additional dosage will be measured as combined chlorine residual. This residual will increase in proportion to the increase in dosage and is identified as zone 2. In this zone, monochloramine is the primary chlorine form.

As the chlorine-to-ammonia molar ratio surpasses 1:1, more and more dichloramine is formed, the dosage/residual relationship will cease to be linear, and the amount of combined and total residual increase will flatten out, as shown by zone 3. In this zone, the proportion of dichloramine increases to a point where it begins to decompose (along with any chloro-organic compounds that may have formed).

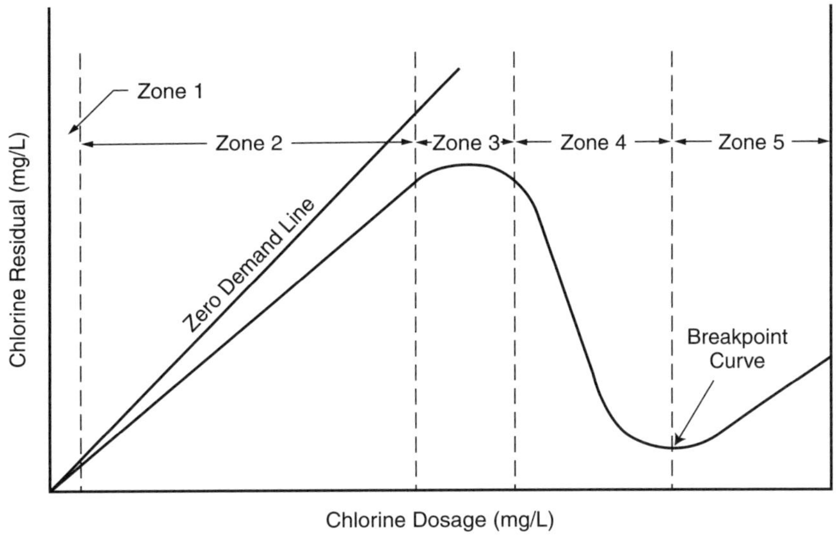

Source: Connell, 1996.

Figure 3-2 Breakpoint curve

Subsequent increases in chlorine dosage will convert an increasing proportion of monochloramine to dichloramine, which subsequently decomposes resulting in a roughly 2-mg/L drop in residual for each milligrams per liter added, as shown by zone 4. This continuing decrease of total residual will taper off and the chlorine residual will reach a minimum point after which it will once again increase with continued chlorine addition. This minimum point is called the *breakpoint* and represents that point in the treatment process at which all ammonia compounds (and most chloro-organic compounds) have been oxidized. Any further chlorine addition will increase the residual chlorine as free chlorine, as identified in zone 5. Combined chlorine residual predominates in zones 2, 3, and 4, while free chlorine residual predominates in zone 5. It is again important to point out that these reactions are not instantaneous. Time must be allowed for the reaction to go to completion before the results follow the breakpoint curve.

Some waters treated by the breakpoint process may never reach a zero residual because organic nitrogen compounds (or other chloro-organic compounds) present resist the oxidation by chlorine. This organic nitrogen is sometimes referred to as an *irreducible minimum*.

Each water plant must determine its own breakpoint curve. Analysis of the water for the various forms of chlorine at different chlorine dosages allows for the determination of the particular breakpoint curve for the water. Procedures from *Standard Methods for the Examination of Water and Wastewater* (APHA, et al., latest edition) should be followed for these determinations.

Inorganic oxidation reactions. The oxidation of soluble iron, manganese, and sulfides is a common use of chlorine in water treatment. Once oxidized iron, manganese, and sulfides form insoluble compounds, they then can be removed by filtration prior to distribution. The following are unbalanced equations that illustrate the oxidation products from chlorination:

$$HOCl + Fe^{2+} \rightarrow Fe^{3+} \text{ iron oxidation} \tag{3-10}$$

$$HOCl + Mn^{2+} \rightarrow Mn^{4+} \text{ manganese oxidation} \tag{3-11}$$

$$HOCl + S^{2-} \rightarrow S \text{ sulfide oxidation} \tag{3-12}$$

$$HOCl + S \rightarrow SO_4^{2-} \text{ sulfur oxidation} \tag{3-13}$$

When iron and manganese are oxidized, they are transformed from a colorless soluble ion to a red and a dark brown precipitate, respectively. If allowed to be oxidized by oxygen or chlorine addition at a customer's tap, the result is the staining of surfaces and cloths and, in some cases, the formation of scales that can eventually restrict flow. Iron is readily oxidized by both free and combined chlorine. Manganese is far more difficult to oxidize (requiring free chlorine, an optimal pH of 10, and 2–3 hr contact time). Approximately 0.6 mg/L chlorine is required to oxidize 1 mg/L of iron. (This reaction consumes 0.9 mg/L of alkalinity.) To oxidize 1 mg/L of manganese requires approximately 1.3 mg/L chlorine (3.4 mg/L alkalinity is consumed). Some organically combined manganese is even more resistant to oxidation.

Sulfide, primarily from wells, is highly toxic and produces a characteristic rotten-egg odor that makes the water unpalatable. With sulfides, there are two reaction steps possible. The first reaction, shown in Eq 3-12, produces colloidal sulfur. The second reaction, shown in Eq 3-13, produces a soluble sulfate ion that has no taste or odor and is not considered objectionable at low concentrations. Which reaction dominates will depend on the pH level and chlorine-to-sulfide ratio. The sulfate-forming reaction will be favored at a low pH (optimum 6.0) and high chlorine-to-sulfide ratio. Sulfur formation is favored at a high pH and low chlorine-to-sulfide ratio. Although oxidation of sulfide is relatively easy, the overall chemistry is complex and its removal without causing objectionable T&O can be difficult (requiring more elaborate treatment processes that are beyond the scope of this discussion).

Organic oxidation reactions. Reactions of chlorine with some organics produce DBPs. The most common are trihalomethanes (THMs). THMs include chloroform, bromodichloromethane, dibromochloromethane, and bromoform. Some of these compounds are classified as probable human carcinogens (Fielding et al. 1993). The current maximum contaminant level (MCL) for THMs in drinking water is 0.080 mg/L (80 µg/L) (USEPA 1996). Another group of DBPs are haloacetic acids (HAAs), which include a number of compounds (11 or more). The five most common HAAs are grouped for regulatory purposes (dichloroacetic acid, monochloroacetic acid, trichloroacetic acid, monobromoacetic acid, and dibromoacetic acid) and the MCL is 0.060 mg/L. Other products of chlorine-organic reactions have also been identified (Kleinjans 1991). Table 3-1 is a list of some of the generally recognized DBPs.

Table 3-1 Chlorination disinfection by-products

Trihalomethanes	Haloacetic Acids
Chloroform	Dichloroacetic acid
Bromodichloromethane	Monochloroacetic acid
Dibromochloromethane	Trichloroacetic acid
Bromoform	Monobromoacetic acid
Haloacetonitrile	Dibromoacetic acid
Dichloroacetonitrile	Tribromoacetic acid
Trichloroacetonitrile	Bromochloroacetic acid
Bromochloroacetonitrile	Bromodichloroacetic acid
Dibromoacetonitrile	Dibromochloroacetic acid
Tribromoacetonitrile	Others
Cyanogen halides	Chloral hydrate
Halopicrins	Haloketones

Source: Fielding et al. 1993.

Many factors influence the formation of DBPs, including contact time (CT), temperature, pH, precursor type and concentration, disinfectant type and concentration, ratio of oxidant to precursor, and concentrations of bromide and ammonia. In addition, naturally occurring humic and fulvic acids are known precursors for DBP formation.

Chlorine can be used as a preoxidant for the removal of T&O and color. Current regulations, however, limit the use of chlorine for this purpose because of the possible high production of undesirable DBPs. Generally, the use of pretreatment methods that remove DBP precursors or inhibit the formation of DBPs is the preferred water treatment strategy. These methods may include one or more of the following: filtration, lime softening, preoxidation with ozone, chlorine dioxide, permanganate, relocation of the point of chlorination, enhanced coagulation, membrane filtration, activated carbon treatment, or ultraviolet (UV) irradiation.

The reaction rates with organic compounds are generally slower than those with inorganic compounds and are often more difficult to quantify because the rate, demand, and end products may vary considerably with reaction conditions and type of precursor. The chlorine requirements are best established by laboratory jar testing to determine the desired treatment level.

Typically, chlorine requirements change periodically in surface water treatment plants. The complexity of new regulations may require pilot plant evaluation to effectively determine the best overall treatment and the most effective role for chlorine. When chlorine is used in the presence of organic compounds, the reaction products are of concern and the results should be reviewed with this in mind.

DISINFECTION MECHANISM

Disinfection is the treatment process used to destroy or inactivate disease-causing (pathogenic) organisms (Von Huben 1995). The consequences of waterborne disease range from mild illness to death. Disinfection should not be confused with sterilization. Sterilization is the complete destruction of all living microorganisms. It has been found that treatment for turbidity removal and subsequent disinfection to control disease-causing organisms is sufficient to protect public health.

Inactivating Pathogens in Water

Most pathogens are accustomed to living in the temperatures and conditions found in the bodies of humans and warm-blooded animals. In general, their ability to survive outside of this environment is limited, but some do survive long enough to cause infections if ingested in drinking water. Certain viruses and protozoans that form cysts can survive for surprisingly long periods, even under adverse conditions. Some pathogenic organisms also tend to be somewhat resistant to certain disinfection processes, so disinfection alone cannot always be assumed to ensure safe drinking water.

Some pathogens can be inactivated by simply storing water in open tanks for extended periods of time. Some pathogens are removed by sedimentation in those tanks, and others experience natural die-off. This is not usually a practical treatment method because of the large investment required for the storage facilities and reliability of the process. In addition, the water is subject to contamination and other nuisance organisms—such as algae—while in storage.

Water disinfection strategies are usually divided into two categories: surface water and groundwater. Surface water is characterized as containing a higher degree of suspended material, higher organic levels, diverse natural and man-made contaminants, variable quality, and subject to a wider variety of pathogens, including most notably *Giardia* and *Cryptosporidium*. Groundwater is generally characterized by low levels of organics and consistent quality. Groundwater is also subject to higher levels and broader types of inorganic compounds, with viruses and bacteria being the pathogens of concern, albeit at low levels.

Surface water treatment is the more involved of the two because the water has not benefited from the natural filtering process of groundwater. Particulate matter (mineral and organic) as well as larger disinfectant-resistant organisms such as *Giardia* and *Cryptosporidium* must be removed by physical separation, primarily sedimentation and filtration. Higher organic levels and oxidation pretreatment requirements also require higher levels and longer contact times for oxidative disinfectants, which may aggravate DBP formation and complicate disinfection. Surface water may also be subject to rapid quality changes. It is not uncommon for surface water plants to use chlorine dioxide, ozone, or permanganate for oxidative pretreatment, followed by free chlorine disinfection and chloramine residual disinfection.

The longer contact time required for inactivation of *Giardia* and the outright resistance of *Cryptosporidium* to inactivation by chlorine present some of the most difficult disinfection challenges to water treatment operators. Treatment is further complicated by the difficulty in testing for these pathogens, making it almost impossible to monitor disinfection performance on a real-time basis. Treatment strategies generally involve a combination of filtration and disinfection designed to achieve a minimum total log reduction.

Disinfection of groundwater is far easier to achieve because water quality changes little over long periods of time. Disinfection programs are generally consistent and, therefore, easier to monitor and control. The pathogens of concern are viruses and bacteria that are relatively easy to inactivate. (Exceptions are in parts of the Southeast and in isolated aquifers elsewhere.) In most groundwater sources, the DBP precursor concentrations are generally low so that in many waters free chlorine can be used without risk of violating DBP regulations. However, it is important to point out that every water system has unique characteristics that must be taken into account to achieve the best water quality.

Detecting Pathogens in Water

Detecting various types of pathogens such as viruses and protozoan cysts is time-consuming and often requires involved analytical techniques. Alternatively, indicator organisms such as fecal coliform and total coliform are used for routine monitoring, and relatively simple, inexpensive tests are available for detecting their presence. These tests, however, only indicate the likelihood that water is contaminated by feces from a warm-blooded animal and, therefore, possibly contains pathogens. All public water systems are required by federal and state regulations to collect representative samples from the distribution system periodically for coliform analysis.

This inability to conduct routine tests for the presence of specific disease-causing microorganisms has been recognized in the US Environmental Protection Agency's (USEPA) Surface Water Treatment Rule, which is discussed later in this chapter. In essence, the rule requires a "treatment technique" for all systems using surface water sources. The technique must consist of one or more methods of treatment that will ensure almost complete removal and/or inactivation of the most resistant pathogenic organisms. In other words, establishing an MCL for pathogenic organisms is not practical because the tests for their presence cannot be completed in a timely manner. Compliance with regulations is based on properly operating the treatment process known to remove or inactivate the organisms.

DISINFECTION METHODS

A 1999 survey revealed the following US disinfection practices in larger water treatment plants (a large majority of smaller water treatment facilities generally use a form of chlorine):

Chlorine gas	83.3 percent*
Sodium hypochlorite	18.3
Sodium hypochlorite (onsite)	2
Chlorine dioxide	8.1
Ozone	5.6
Chloramines	29.4
UV	< 1

*The total percentage is greater than 100 percent because some utilities use several disinfection practices.

Although chlorination is the most common disinfection method, other methods are available and can be used in various situations as described in Table 3-2.

Refer to the *Interim Voluntary Security Guidance for Water Utilities* (AWWA 2004) when evaluating options for disinfection method.

Chemical Treatment

Although the primary use of chemical oxidants is for disinfection, these chemicals also serve other purposes during the disinfection process. In some cases, the choice of chemicals used in a treatment system is dictated by the ability of the chemicals to perform these secondary functions, which include

- control of biological growth in pipelines and basins;
- control of tastes and odors;
- removal of color;
- aids to flocculation; and
- oxidation of iron, manganese, and sulfides.

Aside from oxidizing disinfectants, oxidants such as potassium permanganate and oxygen can be used solely for their oxidation properties.

Table 3-2 Common disinfection methods

Disinfectant Type	Primary Application
Chloramines	Potable water
Iodine	Potable water in emergencies
Bromine	Nondrinking water uses
Chlorine dioxide	Potable water
Ozone	Potable water
Filtration/membranes	Potable water
Ultrasonic/ultrasound	Nondrinking water uses
Ultrahigh frequency	Nondrinking water uses
UV/advanced oxidation processes	Potable water
Ionized radiation (gamma, electron beam)	Nondrinking water uses
Cations of heavy metals (silver)	Nondrinking water uses
Biocides	Nondrinking water uses

Chlorine and Chlorine Compounds

Chlorination is by far the most common form of disinfection currently practiced in the United States. It is the only disinfectant recognized as an acceptable residual disinfectant. When properly understood and correctly operated, the chlorination process is a safe, practical, and effective way to destroy many disease-causing organisms (Figures 3-3 and 3-4).

Principle of Disinfection by Chlorination

The effectiveness of chlorination depends primarily on four factors: concentration (C), contact time (T), pH, and temperature. These factors are combined in a relationship called a CT or concentration (in milligrams per liter) multiplied by the time (in minutes) at a given pH and temperature. CTs have been established for a number of pathogens, disinfectants, and levels of inactivation achieved. (See appendix C.)

The effectiveness of chlorination is also related to the temperature of the water. At lower temperatures, bacterial kill and all chemical reactions in general tend to be slower. However, chlorine is more stable in cold water, and the residual will remain for a longer period of time, compensating to some extent for the lower rate of disinfection. Other factors being equal, all disinfection processes are more effective at higher water temperatures. As temperatures change seasonally, this relationship must be included when adjusting the chlorine dosage.

It is essential that the operator understand and use these relationships in order to obtain the most effective disinfectant. The pH of the water should be checked routinely. This is particularly important if the pH of the water is being raised during disinfection to control corrosion.

It is important to recognize that the use of chlorine as the disinfectant is only one part of the multiple barrier treatment process. In fact, the USEPA acknowledges this in the drinking water regulations under the Safe Drinking Water Act (SDWA). Of equal importance is the need for improved physical removal by sedimentation and filtration. A combination of proper disinfection and improved removal by filtration often provides the most optimum treatment sequence. Chlorination alone, for example, is not recognized as an effective means for *Cryptosporidium* inactivation (Finch 1997).

Residual disinfection. All surface waters and most groundwater systems are required to maintain a measurable residual disinfectant to guard against recontamination, control biofilms and biological growth, meet the USEPA Coliform Rule, and provide an indicator of system integrity and quality. Only chlorine and chloramines are accepted universally as residual disinfectants.

The USEPA's maximum residual disinfection levels (MRDLs) are 4 mg/L for chlorine and chloramines, and 0.8 mg/L for chlorine dioxide.

Interference Substances

Chlorine acts as an effective disinfectant only if it comes in contact with the organisms to be killed. Turbidity, caused by tiny particles of minerals or organic matter suspended in the water, can prevent good contact and protect the pathogens from the effects of chlorine. Therefore, for chlorination to be effective, turbidity must be reduced to very low levels.

As discussed earlier, chlorine reacts with other substances in water, such as organic matter and ammonia. Because these compounds result in the formation of combined residuals, their concentrations are an important factor in determining chlorine dosages.

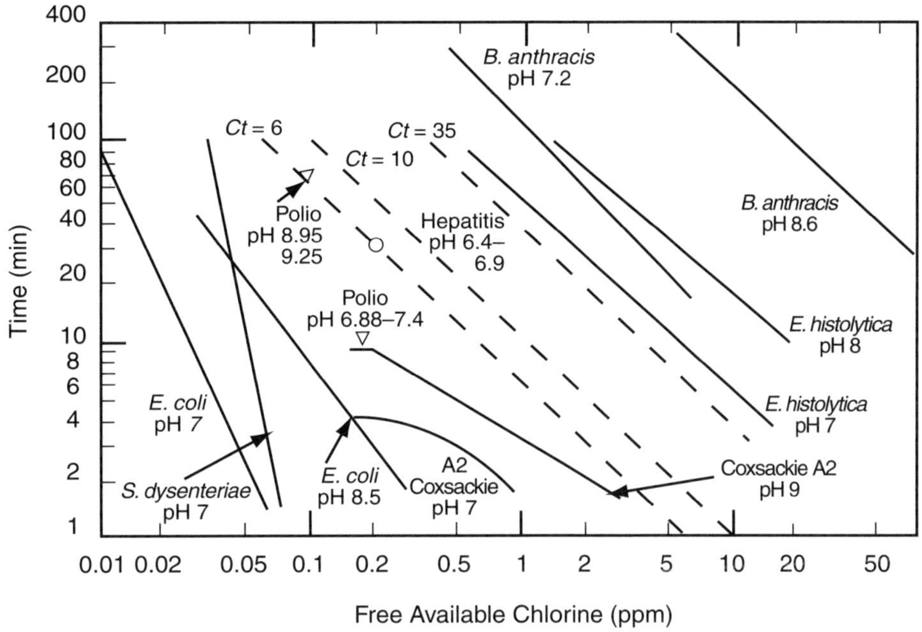

Source: Baumann and Ludwig, 1962.

Figure 3-3 Disinfection versus free available chlorine residuals. Time scale is for 99.6 to 100 percent kill. Temperature was in the range of 20 to 29°C, with pH as indicated.

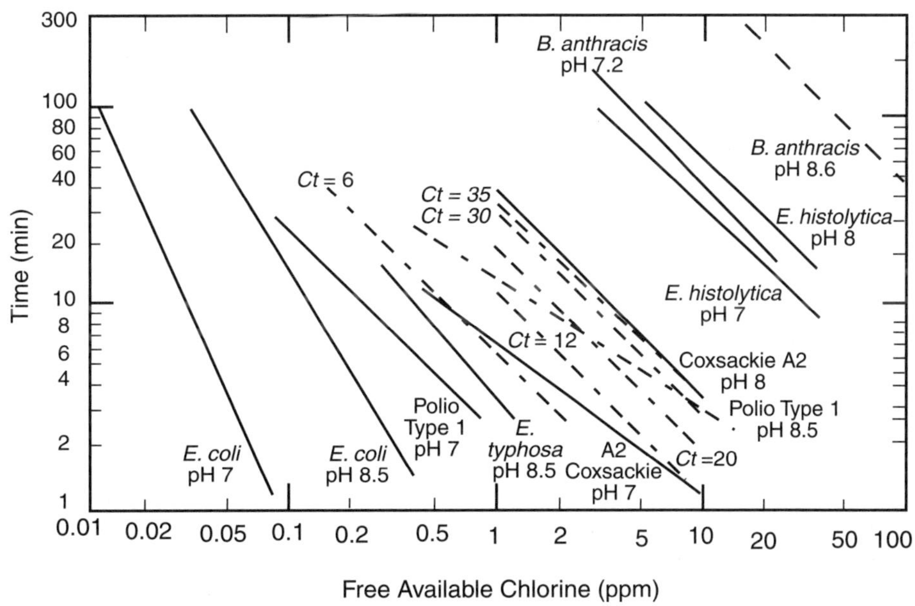

Source: Baumann and Ludwig, 1962.

Figure 3-4 Disinfection versus free available chlorine residuals. Time scale is for 99.6 to 100 percent kill. Temperature was in the range of 0 to 5°C, with pH as indicated.

REFERENCES

APHA, AWWA, and WEF (American Public Health Association, American Water Works Association, and Water Environment Federation). *Standard Methods for the Examination of Water and Wastewater.* Latest edition. Washington, D.C.: APHA.

AWWA (American Water Works Association). 2004. *Interim Voluntary Security Guidance for Water Utilities.* Denver, Colo.: American Water Works Association.

Baumann, E.R., and D. Ludwig. 1962. Free Available Chlorine Residuals for Small Nonpublic Water Suppliers. *Jour. AWWA,* 54:11:1379–1388.

Connell, G.F. 1996. *The Chlorination/Chloramination Handbook.* Denver, Colo.: American Water Works Association.

Fielding, M.N. Mole, P. Jackson, and H. Harth. 1993. *Review of Chlorination and Other Disinfection By-products and Implications for the U.K. Water Industry.* Report UC 1886. United Kingdom: Water Research Centre.

Finch, G.R., L.L. Gyurek, M. Belosovic, and L.R.J. Liyanage. 1997. *Effect of Various Disinfection Methods on the Inactivation of* Cryptosporidium. Denver, Colo.: American Water Works Association and Awwa Research Foundation.

Kleinjans, J.C.S., H.J. Albering, A. Mark, J.M.S. van Maanen, and B. van Angen. 1991. Nitrate Contamination of Drinking Water: Evaluation of Genotoxic Risk in Human Populations. *Environ. Health Perspect.* 94:189–193.

US Environmental Protection Agency. 1996. Safe Drinking Water Act Amendments of 1996. www.epa.gov/safewater/sdwa/summ.html.

Von Huben, H. 1995. *Principles and Practices of Water Supply Operations: Water Treatment, Second Edition.* Denver, Colo.: American Water Works Association.

ADDITIONAL SOURCES OF INFORMATION

AWWA (American Water Works Association). 1999. *Water Quality and Treatment,* 5th ed. R.D. Letterman, ed. New York: McGraw-Hill.

Chlorine Institute, Inc. *Communicating Your Chlorine Risk Management Plan.* Arlington, Va.: The Chlorine Institute Inc.

———. 2001. *Personal Protective Equipment for Chlor-Alkali Chemicals.* Pamphlet 65. Arlington, Va.: The Chlorine Institute, Inc.

Krazner, S.W., and S.E. Barrett. 1984. Aroma and Flavor Characteristics of Free Chlorine and Chloramines. In *Proc. of the AWWA Water Quality Technology Conference.* Denver, Colo.: AWWA.

University of California. 1976. *Kinetics of Breakpoint Chlorination and Disinfection.* Report 76-2. Berkeley, Calif.: Sanitary Engineering Research Laboratory, University of California.

Chapter 4

Chlorine and Ammonia: Handling, Storage, Feed Equipment, and Systems

GAS CHLORINATION FACILITIES

Chlorine gas (Cl_2) is approximately 2.5 times as dense as air (Von Huben 1995). It has a pungent, noxious odor and a greenish-yellow color, although it is visible only at a very high concentration. Chlorine gas is irritating to the eyes, nasal passages, and the respiratory tract and it can kill a person in a few breaths at concentrations as low as 0.1 percent (1,000 ppm [mg/L]) by volume. Its odor can be detected at concentrations above 0.3 ppm.

Chlorine liquid, which is approximately 99.5 percent pure chlorine, is created by compressing chlorine gas. The liquid is amber in color and approximately 1.5 times as dense as water. It can be purchased in cylinders, containers, tank trucks, and railroad cars, as shown in Figures 4-1 through 4-4.

Liquid chlorine changes easily to a gas at room temperatures and pressures. One volume of liquid chlorine will expand to approximately 460 volumes of gas. Dry chlorine gas will not corrode steel or other metals, but it is extremely corrosive to most metals in the presence of moisture.

Chlorine will not burn. But, like oxygen, *it will support combustion*—that is, it takes the place of oxygen in the burning of combustible materials. Chlorine is not explosive, but it will react violently with greases, turpentine, ammonia, hydrocarbons, metal filings, and other flammable materials. Chlorine will not conduct electricity, but the gas can be very corrosive to exposed electrical equipment. Because of the inherent hazards involved, chlorine requires special care in storage and handling.

HANDLING AND STORING CHLORINE GAS

Safe handling and storage of chlorine are vitally important to the operator and to the communities immediately surrounding a treatment plant, because an error or accident in chlorine handling can cause serious injuries or even fatalities. The containers commonly used to supply chlorine in smaller water treatment plants are 150-lb (68-kg) cylinders. Larger plants find it more economical to use ton containers. Some large plants are equipped to draw chlorine directly from tank cars.

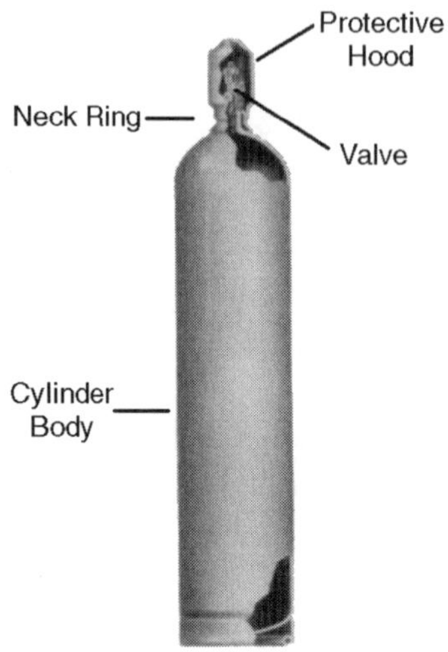

Courtesy of the Chlorine Institute, Inc.

Figure 4-1 Chlorine cylinder

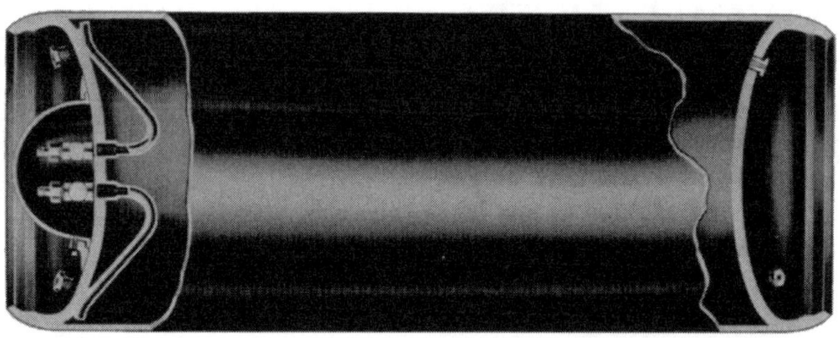

Courtesy of the Chlorine Institute, Inc.

Figure 4-2 Chlorine ton container

Courtesy of PPG Industries, Inc.

Figure 4-3 Chlorine ton container truck

Courtesy of the Chlorine Institute, Inc.

Figure 4-4 Chlorine tank car

The decision of whether to use cylinders or ton containers should be based on cost and capacity. The cost per pound (kilogram) of chlorine in cylinders is usually substantially more than that of chlorine in ton containers. If a plant's needs for chlorine are lower than 50 lb/d (23 kg/d), cylinders should usually be selected. For systems that use large amounts, ton containers will probably be more economical.

Cylinders

Chlorine cylinders hold 150 lb (68 kg) of chlorine and have a total filled weight of 250–285 lb (110–130 kg). They are approximately 10.5 in. (270 mm) in diameter and 56 in. (1.42 m) high. As illustrated in Figure 4-1, each cylinder is equipped with a hood that protects the cylinder valve from damage during shipping and handling. The hood should be properly screwed in place whenever a cylinder is handled, and should be removed only during use.

Cylinders are usually delivered by truck. Each cylinder should be unloaded to a dock at truck-bed height if possible. If a hydraulic tailgate is used, the cylinders should be secure to keep them from falling.

Cylinders must never be dropped, including "empty" cylinders, which actually still contain some chlorine. The easiest and safest way to move cylinders in the plant is with a hand truck. The hand truck should be equipped with a restraining chain that fastens snugly around the cylinder about two thirds of the way up. Slings should never be used to lift cylinders, and a cylinder should never be lifted by the protective hood because the hood is not designed to support the weight of the cylinder.

Cylinders should not be rolled to move them about a plant. Tipping the cylinders over and standing them up can lead to employee injury. In addition, the rolled cylinders might strike something that could break off the valve.

Cylinders can be stored indoors or outdoors. If cylinders are stored indoors, the building should be fire resistant, have multiple exits with outward-opening doors, and be adequately ventilated. Outdoor storage areas must be fenced and protected from direct sunlight, and they should be protected from vehicles or falling objects that might strike the cylinders. If standing water accumulates in an outdoor storage area, the cylinders should be stored on elevated racks. Avoiding contact with water will help minimize cylinder corrosion.

Some operators find it convenient to hang "full" or "empty" identification tags on cylinders in storage, so that the status of the chlorine inventory can be quickly determined. Other plants maintain separate storage areas for full and empty cylinders, but all cylinders, full or empty, should receive the same high level of care. In addition, protective hoods should be placed on empty and full cylinders in storage. Even when a cylinder no longer has sufficient chlorine for plant use, a small amount of gas remains and could escape if the cylinder or valve was damaged. Both full and empty cylinders should always be stored upright and secured with a chain to prevent them from tipping over.

Ton Containers

The ton container is a reusable, welded tank that holds 2,000 lb (910 kg) of chlorine. Containers weigh about 3,700 lb (1,700 kg) when full and are generally 30 in. (0.76 m) in diameter and 80 in. (2.03 m) long. As shown in Figure 4-2, the ends are concave. The container is crimped around the perimeter of the ends, forming good gripping edges for the hoists used to lift and move them. The ton container is designed to rest horizontally both in shipping and in use. It is equipped with two valves that provide the option of withdrawing either liquid or gaseous chlorine. The upper valve draws gas, and the lower valve draws liquid.

Handling the heavy containers is, by necessity, far more mechanized than handling cylinders. Containers are loaded or unloaded by a lifting beam in combination with a manual or motor-operated hoist mounted on a monorail that has a capacity of at least 2 tons (1,815 kg), as shown in Figure 4-5. To prevent accidental rolling, containers are stored on trunnions, as illustrated in Figure 4-6. The trunnions enable the container to be rotated so that it can be positioned correctly for connection to the chlorine supply line.

Ton containers can be stored indoors or outdoors and require the same precautions as chlorine cylinders. The bowl-shaped hood that covers the two valve assemblies when the tank is delivered should be replaced each time the container is handled, as well as right after it has been emptied.

The chlorine storage area should provide space for a 30- to 60-day supply of chlorine. Some systems feed chlorine directly from this storage area. When ton containers are used, the chlorination feed equipment is usually housed in a separate room, as shown in Figure 4-7.

Figure 4-5 Lifting beam with motorized hoist for ton containers

Figure 4-6 Ton containers stored on trunnions

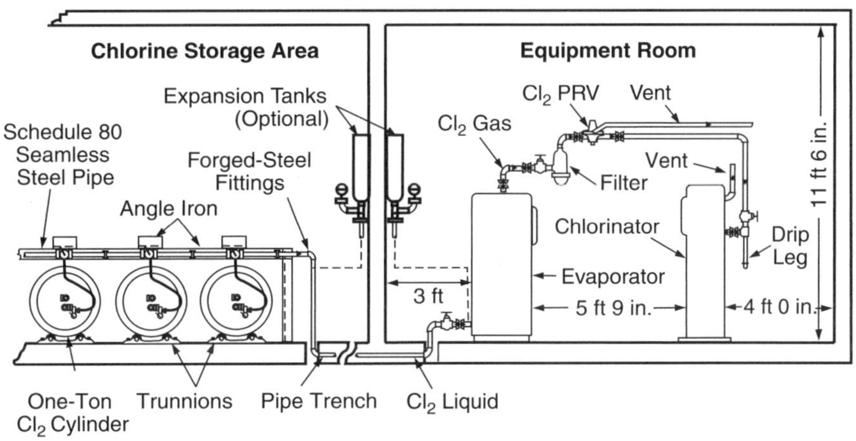

Source: Handbook of Chlorination by Geo. Clifford White, copyright © 1972 by Van Nostrand Company. Reprinted by permission of the publisher.

Figure 4-7 Chlorination feed equipment located in a separate room

Tank Cars/Rail Cars

Chlorine is available in tank cars and rail cars for very high capacity needs. Semi-trailer tank cars come in various sizes but many have a capacity of 25 tons. Rail cars also vary in size but a common capacity is 90 tons. These large storage containers require special unloading systems that consist of the following:

- unloading platform,
- storage tank and sun shield,
- weighing device,
- air padding system,
- eductor,

- chlorine gas and liquid headers,
- gauges,
- pressure switches and alarms,
- expansion tanks,
- flexible connections, and
- spill containment and neutralization system.

FEEDING CHLORINE GAS

Chlorine feeding begins where the cylinder or ton container connects to the manifold that leads to the chlorinator. The feed system ends at the point where the chlorine solution mixes into the water being disinfected. The main components of the system are the following:

- weighing scale,
- valves and piping,
- chlorinator, and
- injector and diffuser.

Weighing Scales

It is important that an accurate record be kept of the amount of chlorine used and the amount of chlorine remaining in a cylinder or container. A simple way to do this is to place the cylinders or containers on weigh scales. The scales can be calibrated to display either the amount used or the amount remaining. By recording weight readings at regular intervals, the operator can develop a record of chlorine-use rates. Figure 4-8 illustrates a common type of two-cylinder scale. Figure 4-9 illustrates a portable beam scale. Figure 4-10 illustrates a combination trunnion and scale for a ton container, which operates hydraulically and has a dial readout.

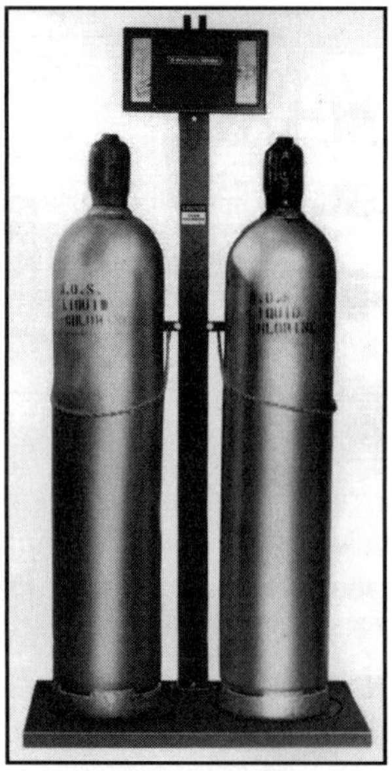

Courtesy of Wallace & Tiernan, Inc.

Figure 4-8 Two-cylinder scale

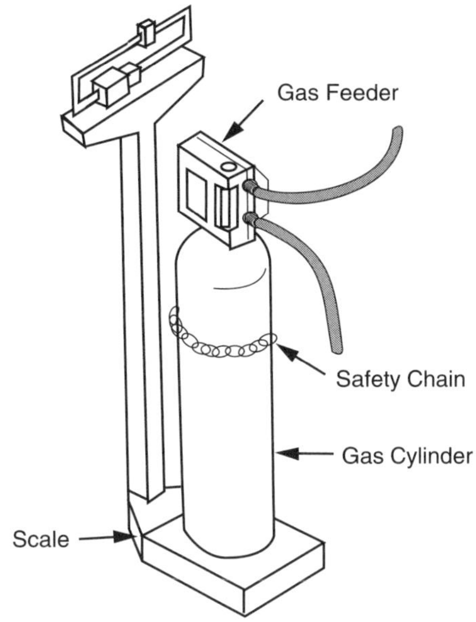

Courtesy of Capital Controls Company, Inc.

Figure 4-9 Portable beam scale

Courtesy of Force Flow Equipment.

Figure 4-10 Combination trunnion and scale for a ton container

Valves and Piping

Chlorine cylinders and ton containers are equipped with valves as shown in Figures 4-11 and 4-12. The valves must comply with standards set by the Chlorine Institute (Chlorine Institute 1998a).

It is standard practice for an auxiliary tank valve to be connected directly to the cylinder or container valve, as illustrated in Figure 4-13. The connection is made with either a union-type or yoke-type connector. The auxiliary valve can be used to close off all downstream piping, thus minimizing gas leakage during container changes. The auxiliary tank valve will also serve as an emergency shutoff if the container valve fails. If a direct-mounted chlorinator is used, an auxiliary tank valve is not required.

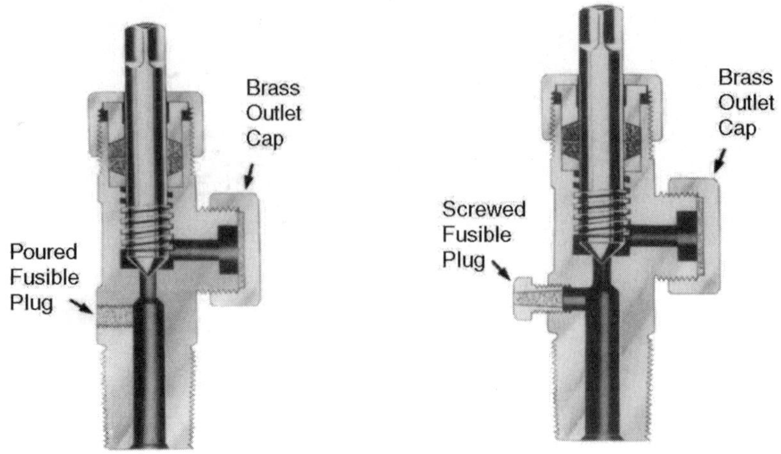

Courtesy of the Chlorine Institute, Inc.

Figure 4-11 Standard cylinder valves: poured-type fusible plug (left) and screw-type fusible plug (right)

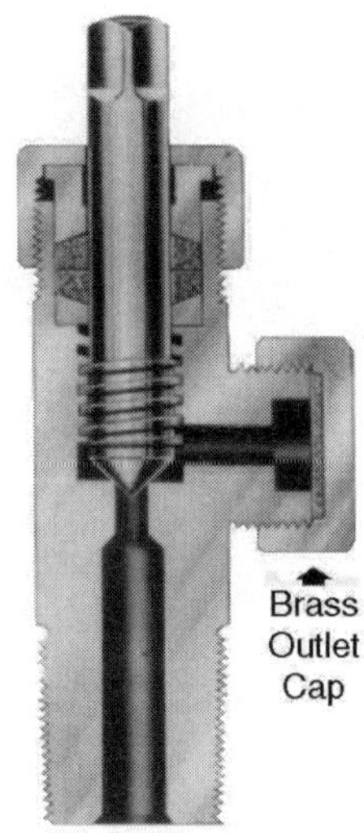

Courtesy of the Chlorine Institute, Inc.

Figure 4-12 Standard ton container valve

The diagram in Figure 4-13 is of a typical valve assembly. The figure shows that the assembly is connected to the chlorine-supply piping by flexible tubing, which is usually 3/8-in. (10-mm) copper rated for 500 psig (3,500 kPa).

When more than one container is connected, a manifold must be used. The manifold channels the flow of chlorine from two or more containers into the chlorine-supply piping. The manifold and supply piping must meet the specifications of the Chlorine Institute. Manifolds may have from 2 to 10 connecting points. Each point is a union nut suitable for receiving flexible connections. Notice that the header valve is connected at the manifold discharge end, providing another shutoff point. Additional valves are used along the chlorine supply line for shutoff and isolation in the event of a leak.

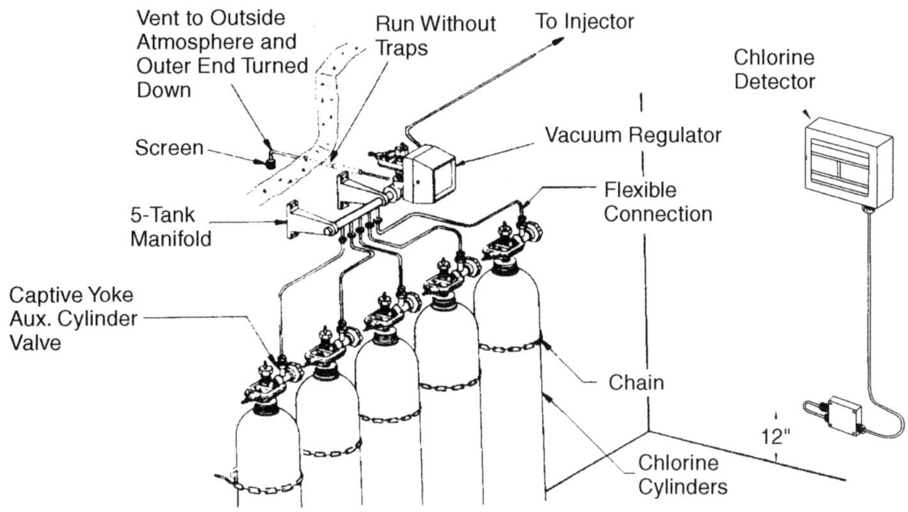

Courtesy of US Filter / Wallace & Tiernan.

Figure 4-13 Auxiliary tank valve connected directly to container valve

Chlorinators

Equipment to feed chlorine gas is designed to work either under pressure or under vacuum (Connell 1996). By far the most common of the two types operates under vacuum called the *vacuum-operated solution feed chlorinator*. The name stems from the fact that the equipment feeds chlorine gas only when it receives a vacuum signal, and the gas that is fed is mixed with water to form a highly concentrated solution fed to the point of application. The other type of chlorine gas feeder, pressure-operated gas feed, operates under the pressure supplied by the gas and feeds gas to the point of application.

The vacuum-operated units offer greater safety in the operation of the equipment and handling of chlorine gas. Such units also provide for greater versatility in the application and control of the dosage. The principal components of a chlorine gas feeder are vacuum regulators, flow indicators, flow controllers, and a venturi. The feeder operates by regulating the flow of chlorine gas by controlling and regulating the vacuum conditions upstream and downstream of an orifice or flow-control device.

Most chlorine gas feeders use two methods of controlling and regulating the gas flow: constant differential-pressure or sonic flow. Constant differential requires maintaining a constant vacuum differential across the orifice (rate-control valve) by using a differential pressure (vacuum) regulator or downstream vacuum regulator. This regulator maintains a constant pressure (vacuum) drop across the orifice for any given setting of the orifice. When this occurs, the operator can adjust the orifice (the rate-control valve) and be assured that the gas flow will be maintained at the desired setting. The use of a differential-pressure regulator or a downstream vacuum regulator corrects for any variation in downstream vacuum that would cause an undesirable variation in gas flow.

Sonic flow requires maintaining a minimum upstream vacuum so that a variation in downstream vacuum will not cause any variation in flow. Sonic flow is so named because the minimum upstream vacuum level required to reach the desired condition is such that the velocity of the gas through the orifice is at the speed of sound, i.e., sonic velocity.

The ratio of the downstream pressure to the upstream, for which the sonic velocity is attained, is called the *critical pressure ratio* (rc). If the upstream vacuum is held at a constant 20 in. (508 mm) water (13.973 psia [96.3 kPa (absolute)]), the downstream vacuum must be 14.6 in. (371 mm) mercury vacuum or 7.53 psia (51.9 kPa [absolute]).

The flow of chlorine gas across an orifice increases as the ratio (P_u:P_d) decreases. Sonic flow of chlorine is reached when this ratio reaches 0.539. When $P_u = P_d$, the ratio is 1 and flow ceases. As the vacuum level increases, the gas flow increases, although not linearly. At some point, any additional increase in vacuum level causes no further increase in gas flow. The point at which this occurs is the sonic flow of the gas. For chlorine, sonic flow occurs at approximately 14 in. (356 mm) of mercury vacuum level downstream of the orifice. Sonic flow is important because it represents a simpler form of flow control, has fewer components than other typical functioning systems, and reduces the costs of maintenance and service.

Induction mixers. Induction mixers (Figure 4-14) can be used in feeding gases, such as chlorine or ammonia, and solutions, such as sodium hypochlorite (NaOCl), aqueous ammonia, or other ammonia salts used in water treatment. These devices enhance the safety of the system design by extending the vacuum line into the mixing chamber and by eliminating the need for external mixers. By eliminating solution lines, the installation has been simplified and a more rapid response to a feed change is possible. Chemical savings of 20–40 percent have been identified by one manufacturer. These savings depend on the process and installation needs and thus are site specific. The use of induction mixers has been extended to other chemicals used in water treatment, for example, carbon dioxide and ferric chloride.

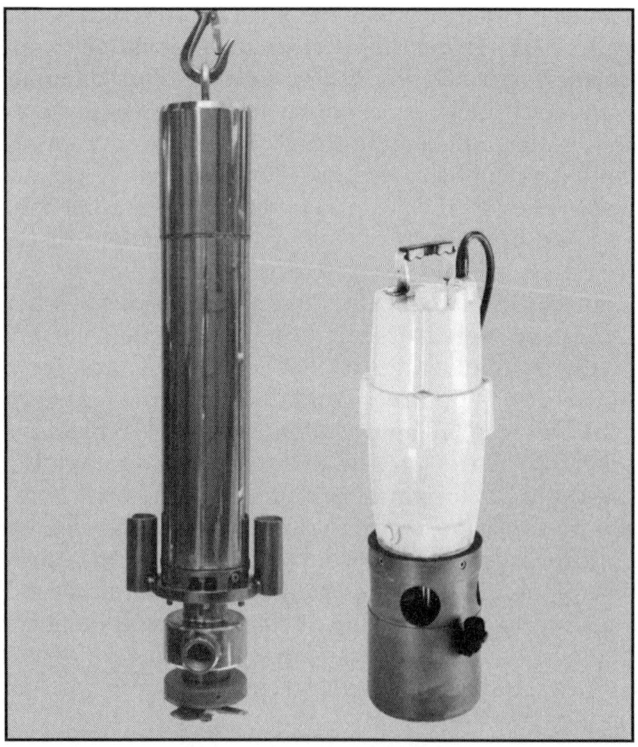

Source: Gardiner Equipment Company, Inc. (left); Capital Controls Company, Inc. (right).

Figure 4-14 Chemical induction mixers use a high-speed impeller to create a vacuum to draw the chemical, gas, or liquid into intimate contact with water to be treated. Some mixers have an open impeller design (left) and some are closed (right).

Materials of construction. The materials used in the construction of gas chlorinators are chosen specifically to handle chlorine gas. The materials are not designed to handle liquid chlorine (Chlorine Institute 1999). All attempts must be made to keep liquid chlorine from entering a gas chlorinator.

Major structural components of chlorinators are normally made of acrylonitrile butadiene styrene (ABS) or polyvinyl chloride (PVC). Some components may be made of chlorinated polyvinyl chloride (CPVC), which provides some additional features, such as higher temperature resistance. Regulator diaphragms are constructed of polychlorotrifluoroethylene (PCTFE), polytetrafluoroethylene (PTFE), or similar fluorocarbons; sealing O-rings of fluorocarbon rubber or similar material; shutoff valves of PTFE and silver; and regulator springs of Tantalloy or Hastelloy C.

All materials must be resistant to corrosion from wet chlorine because there is a possibility that moisture can enter the chlorinator through the ejector check valve and form corrosive acids. Thus, the use of silver, Tantalloy, and similar corrosion-resistant metals is preferable. The noble metals (silver and gold) are corrosion resistant to both wet and dry chlorine.

Piping for chlorine gas under vacuum up to 6 psig (41.3 kPa [gauge]) can be of plastic construction, such as PVC, polyethylene (PE), or similar materials (Chlorine Institute 1998a). Steel pipe, satisfactory for use in dry pressure lines, is not recommended for chlorinator vacuum lines where the presence of moisture can occur if the ejector check valve leaks. If steel pipe is desired for chlorinator vacuum lines, it must be lined with a corrosion-resistant coating. These linings are generally made of PVC or other elastomeric material.

Reliquefaction. As discussed in chapter 5, whenever a chlorine gas feeder installation makes use of gas pressure piping to connect the feeding equipment with the source of chlorine, reliquefaction (the formation of chlorine droplets) in the pressure piping can occur. This is to be avoided at all costs because the gas feeder is designed to handle gas and not liquid chlorine or chlorine droplets.

Reliquefaction will occur when the gas is saturated or at a temperature and pressure such that its physical condition is located at a point on the vapor pressure curve shown in Figure 2-1 in chapter 2. Any reduction in temperature will cause condensation to occur on the pipe surface. This condensation can be carried as liquid droplets along with the gas stream to the gas feeder. Following are some techniques that may help avoid reliquefaction:

- Piping should never pass from a warmer area to a cooler area. The ambient temperature should always increase in each successive room or area through which the pressure piping passes.
- Piping should always slope toward the source so that any reliquefied chlorine can drain back toward the supply rather than toward the gas feeder. The addition or inclusion of drip legs (down pipes) at points of direction change will help collect any droplets, particularly if the drip leg is equipped with a small (25- to 50-watt) heater.
- Some installations may find the use of a pressure-reducing valve helpful because a drop in pressure will lower the temperature at which reliquefaction occurs.

The most important recommendation is to install the vacuum regulator at the gas-pressure source or as soon as possible after the line exits the container to reduce the gas to a vacuum. When the gas is under vacuum, reliquefaction will not occur until the temperature is below –30°F (–34.4°C). The use of a vacuum regulator mounted directly on a cylinder or ton container valve further reduces the reliquefaction potential because gas pressure lines in the system are eliminated.

The use of flexible connectors to connect the container to a pressure manifold is frequently practiced. When the gas exits the container under pressure and then enters a pressure manifold, a pressure-reducing valve is recommended as soon as possible in the pressure line in order to minimize reliquefaction. However, reliquefaction can still occur at ambient temperatures. For example, if the container and gas pressure line were 80 psig (551.6 kPa [gauge]), the reliquefaction temperature is 67°F (19.4°C). If the pressure was reduced to 65 psig (448.1 kPa [gauge]), liquefaction would occur at 56°F (13.3°C). Refer to Figure 2-1.

Service and maintenance. Each chlorinator manufacturer provides service and maintenance instructions that should be followed for optimum operation. These instructions may vary from manufacturer to manufacturer. However, several fundamental points are common to all manufacturers' manuals. Among these are the following:

- Keep all moisture away from any chlorine system. Any moisture will form acids with chlorine that can cause corrosion, particularly in steel pressure piping. The corrosion products can build up in the piping system and plug the piping so that the flow of gas is severely restricted or stopped. The corrosion products can be carried along with the gas stream to the gas chlorinator. The presence of these materials in the chlorinator will cause blockage in valves and small orifices that will stop gas feed and cause downtime.
- Impurities in chlorine can create feed problems. These impurities may be organic or inorganic in nature. A visual indication of a developing problem can be seen at the gas-metering tube. The tube will darken with brownish-red impurities, such as ferric (iron) chlorides from chlorine reactions with the metallic piping, or an off-white, wax-like material, which consists of chloro-organic compounds left in chlorine from the chlorine manufacturing process. In either case, the chlorinator will have to be taken off line and cleaned. Hot water is usually sufficient to dissolve the iron salts, and an organic solvent such as methanol will usually remove the organics. (Refer to the manufacturer's instruction manual for further details relative to the particular equipment in use.) All impurities tend to collect at points of pressure drop, such as valves and flowmeters. These points must be examined and thoroughly cleaned. Be sure to dry the cleaned components thoroughly before reuse.
- At points of pressure connection, such as at the exit from cylinders or ton containers, it is important to use gaskets that are not brittle. Although frequently found in chlorination installations, fiber gaskets are *not* recommended because they become friable and frequently fray, break off, or disintegrate. The fibers from this type of gasket often find their way into the gas stream and cause blockage or create flow restrictions in the gas chlorinator. Lead gaskets are recommended at frequently changed connections to provide a good seal with a material that has more desirable physical characteristics. These gaskets must be replaced each time the connection is opened.

The life of a flexible connector, sometimes called a pigtail, can be extended by using isolating valves at locations where frequent connection changes are made. Isolating valves protect the open end of the flexible connector from moisture intrusion during the change. The frequency of container changes and moisture content of the chlorine gas affects the life of the flexible connector. Flexible connectors are usually made of copper tubing with external corrosion protection, usually cadmium plating. Copper tubing is quite satisfactory for this purpose. Internal corrosion is possible in the presence of moisture, but often is not externally visible. However, internal corrosion can be detected by listening for squeaking sounds from tubing movement during container connection changes. When squeaking occurs, replace the connectors. Connectors should be changed once a year or more frequently depending on the number of times containers are changed. Because the flexible connector is under pressure, any indication of a leak must be given prompt attention. Flexible connectors that have been crimped must be replaced.

Each manufacturer recommends the necessary spare parts to keep equipment operational and recommends an adequate supply of these spare parts, which should be maintained in an appropriate location. Operating and maintenance personnel should ensure that the supply is readily available and avoid the indiscriminate use of a spare or standby feeder as a source for the parts.

Control of gas feeding equipment. It is important to review the different methods of control for gas feeders, the applications that are more appropriate for the particular method, and an explanation of why these applications are appropriate (Connell 1996). The equipment involved in both manual and automatic control operates by the same principles, although modifications must be added to permit automatic control.

The biggest modification is in the controller. The controller is an electronic instrument that receives the signal or signals from the sensing instruments. The signals are electronically multiplied or combined to provide one signal to the feeding equipment. Electronic control is the industry standard in modern water treatment plants. As recently as the early 1970s, many plants were still equipped with pneumatic controls. Because of their simplicity, economy, and proven performance, electronic control schemes have replaced most (if not all) of the outmoded pneumatic controllers.

This section does not discuss the various controllers available but instead reviews the equipment used for feeding the chemicals, including manual, semiautomatic, automatic proportioning, automatic residual, and compound-loop control gas feeding equipment.

Manual control gas feeder. Manual control indicates that startup, shutdown, and feed-rate adjustments are all done by hand by the treatment plant operator. Startup is accomplished by opening a water valve to the venturi in the injector or starting up a water pump to feed the venturi. The cylinder valve must also be opened to permit gas to flow. The gas feed rate is adjusted to the desired value by turning the rate valve and adjusting the rate by observing the gas flowmeter or indicator. The gas flowmeter will read in pounds per 24-hr day or in grams or kilograms per hour. The amount of gas to be fed is determined from the dosage desired and the water flow rate. Once the feed rate is established, the equipment operates continuously at that rate until the feeder is either shut down or the feed is adjusted to another rate.

Applications appropriate for manual control include situations in which water flow and demand are constant and an operator is readily available to make any necessary adjustments. A typical application of this type is in the treatment of water in swimming pools. This type of application should never be left unattended.

Semiautomatic control gas feeder. Semiautomatic control is sometimes referred to as *start/stop* or *on/off control*. In this control scheme, the operator sets the desired chlorine feed rate manually at a fixed value. Startup and shutdown are done automatically. An electrical signal is supplied from an external source (e.g., a well-pump starter) and is used to activate the gas feeder by opening a solenoid valve and/or starting a booster pump in the water supply line to the venturi in the chlorine injector.

Applications appropriate for semiautomatic control are water systems pumping from underground wells. There is usually no additional treatment of the water provided, nor needed. The water has a fairly constant demand. Semiautomatic control is perhaps the most widespread use for chlorination equipment due to the simplicity of the system needs, the safety of the installation, and the ease of adaptability to water systems.

Automatic proportioning control gas feeder. Sometimes referred to as flow proportioning control, this type of control may be the most frequently used for automatic operation. In instrumentation terminology, flow proportioning control may more properly be called *open-* or *single-loop control*. In open-loop control, there is no signal or feedback to the control system, hence the open loop. This is sometimes referred to as a *wild stream control*.

In flow proportioning control, two additional items are added to the chlorination system. These are a motorized valve and an electronic controller. The valve used is a linearized valve set so that each incremental change provided as input results in a constantly proportioned change in the valve opening. The chlorination equipment is equipped with an electronic controller that receives information from the sensing instrument (water flowmeter), sends the information into the electronic controller, calculates the desired output based on the desired chlorine dosage, and sends a signal to the valve drive motor to open or close the valve the correct amount. The valve must be equipped with a drive device that will open or close the valve as instructed by the automatic controller. The flow signals are usually electronic (4–20 milliamperes, direct current [mADC], or 1–5 volts, direct current [VDC]), and the controller provides the appropriate electrical output to position the valve in direct proportion to the flow signal, hence the name flow proportioning control.

Because the chlorine gas flow is altered only in response to variations in water flow, the chlorine demand in the water to be treated must be constant. The dosage desired by the operator must be set manually to meet the desired residual, and the results must be measured manually or indicated continuously with a continuous chlorine residual analyzer. There is no feedback to the control system, so the operator must be alert to any deviations from the desired result and respond promptly to make the corrections manually.

Applications for flow proportioning control must be (1) where the demand is constant and only the flow will vary or (2) where deliberate, periodic operator input is desired as a part of routine operation to check residual conditions. The water system must have a means of measuring the flow and providing the 4- to 20-mA electrical signal to the gas feeder's controller. Flow proportioning control is applicable for both groundwater systems and waste treatment systems that have varying flows and fairly constant demand.

Automatic residual control gas feeder. Where flow proportioning control fails by providing no feedback to the controller, residual control fills the gap. Residual control is often called *closed-loop control* because there is feedback to the controller of the result of the change, and, thus, the loop is closed. Automatic residual control requires the use of an instrument external to the gas feeder—a chlorine residual analyzer—to provide a means of continuously measuring the residual chlorine in the water to be treated and sending an electrical signal (4–20 mA) to an automatic controller. The controller compares the signal (the residual) to the desired residual, or set point, and varies the valve position or valve opening to feed the desired amount of chlorine to reach the desired residual.

The difference between flow proportioning control and residual control is that residual control includes feedback of the result of the input. Although this scheme provides the result of the input, it does not recognize the source of the change. Changes in water flow, which will cause a change in residual, typically occur rather rapidly. The residual control scheme is not capable of handling a rapid change of flow, and its usefulness under these conditions is limited.

Applications for residual control equipment include situations in which the flow does not vary or varies so slowly that the control scheme can recognize the change and respond to make proper corrections to the chlorine feed rate. Water treatment plants with clearwells of sufficient size may be such an application because the demand is fairly constant and the clearwell can provide sufficient holdup to attenuate flow variations. However, care must be taken with the use of residual control to ensure proper mixing and treatment time. The sampling point is critical to accurate control because the treatment time after chlorine addition is a factor in the process but is not measured as an element in the residual control loop.

For practical purposes, the times in the loop that are electrical in nature are insignificant compared to the other factors. The remaining critical times are of importance and their influence on the control result are minimized if

- the control valve is located as close as possible to the venturi,
- the venturi is as close as possible to the point of application,
- the sample point location is as close to the point of application as the process will permit,
- the sample transfer time to the analyzer is minimized by locating the analyzer as close as possible to the sampling point, and
- the analyzer reads the residual value constantly, not intermittently.

Some European utilities, particularly in the United Kingdom, practice triplication by using three sensors (analyzers). This practice is particularly helpful at unmanned sites because three values provide some assurance of proper analyzer performance. The theory of this practice is that if at least two of the analyzers are reading the same, operating parameters are being met. The third analyzer could be checked and calibrated at a convenient time. This way the correct signal is obtained from among the three analyzers and sent to the controller. The use of three analyzers can be adapted for monitoring as well as control conditions.

Compound-loop control gas feeder. The ideal control scheme can be obtained by taking both the open- or single-loop control and closed-loop control systems and combining them into one scheme, hence the name *compound-loop control*. In this control system, two signals, flow and residual, are sent to one controller that combines the two inputs mathematically. The controller compares the resultant to a set point and sends one electrical signal to the automatic valve. The flow signal is the major control input because water flow changes are instantly measured and the flow signal is corrected rapidly. Meanwhile, the residual signal is used as a final trim for the valve. The advantage of this control scheme is that it combines the features of flow proportioning control (quick response to changes in flow) and residual control (recognizing residual deviations) and adjusts for both at the same time.

Two sensing instruments, the chlorine residual analyzer and the water flowmeter, provide the inputs to the controller. The controller provides the ability to control to a fixed residual in a variable flow and variable demand condition. The location of the sampling point is critical to optimum operation. This sampling point need not be after the required contact time but located within a reasonable time after chemical addition. The decrease in residual is a function of the reaction rate and demand. The controller design must provide for lag-time considerations and be able to compensate for variations in flow that will alter contact times for fixed sampling locations. The ideal controller design should have a feature that permits either input to be used independently for valve control in the event of a failure of the flowmeter or residual analyzer.

Compound-loop control may be used in distribution systems where rechlorination is required and at some water treatment plants where flow and residual may vary. Response to rapid flow changes is a hallmark of these systems, as shown in Figure 4-15.

Injectors and Diffusers

The injector (ejector or eductor) is located on the chlorinator discharge line. A venturi creates a vacuum that pulls chlorine gas from the chlorinator into a stream of dilution water. This results in a strong solution of chlorine and water that is conveyed to the point of application. The strong chlorine solution is corrosive, as the pH is generally 2–4, and, therefore, chemically resistant material must be used (e.g., PVC, fiberglass, steel-lined with PVC or rubber). A strainer is used upstream of the injector to reduce the possibility of clogging due to foreign material blocking the injector throat.

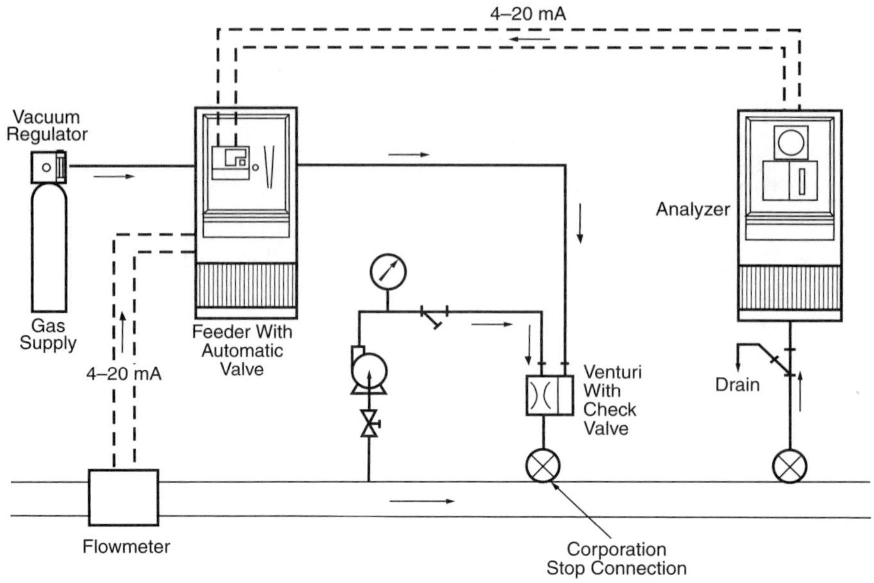

Source: Connell, 1996.

Figure 4-15 Compound-loop control

A diffuser is one or more short lengths of pipe, usually perforated, that distribute the chlorine solution into the main water flow. There are many different types of diffusers (and mixers). The best diffusers accomplish an even distribution of the chlorine solution and promote rapid and complete mixing.

Diffusers are used in two general application situations: pipelines and open channels or tanks. Common diffusers for smaller pipelines (less that 3 ft [0.9 m] in diameter) are simply a pipe (usually PVC) protruding into the center of the pipeline. Effective mixing occurs within 10 pipe diameters from the point of application. For larger pipelines, a perforated pipe diffuser is often employed, as shown in Figure 4-16.

Gas Chlorination Feed System Problems

The following are two of the most common gas chlorination problems:

- Chlorine leaks
- Stiff container valves

Proper maintenance can prevent many problems. Most manufacturers have equipment troubleshooting guides that help locate and correct problems.

Chlorine leaks. A major concern in the operation and maintenance of the chlorination process is the prevention of chlorine leaks (Chlorine Institute 2000a). The most common place for leaks to occur is the pressurized chlorine supply line between the containers and the chlorinator. Every joint, valve, fitting, and gauge in the line is a possible point of leakage.

Some chlorine leaks are readily apparent. Others are slow, small, partly hidden, or otherwise difficult to locate. The usual method of detection is to open a bottle of ammonia solution, place the bottle near a suspected leak, and allow the fumes to rise around the suspected area. If there is a sizable leak, the chlorine will combine with the ammonia to form a visible white vapor. Unfortunately, this method will not indicate a very small leak. In addition, very small leaks will often not produce a noticeable odor. Small leaks can go unnoticed for weeks unless the operator periodically looks for two signs: joint discoloration and moisture.

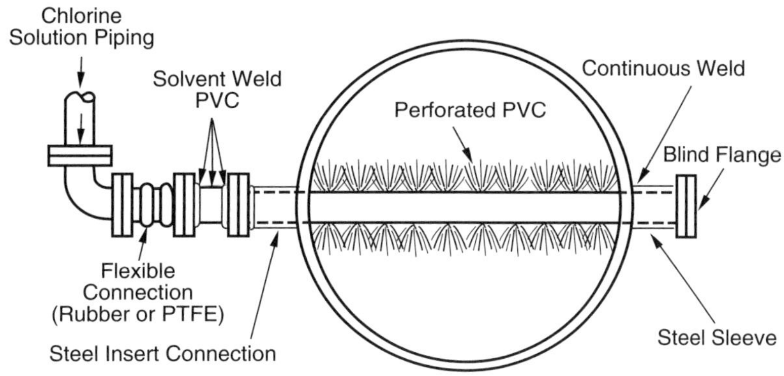

Source: Handbook of Chlorination and Alternative Disinfectants. 4th ed. by Geo. Clifford White, copyright © 1998. Reprinted by permission of John Wiley & Sons, Inc.

Figure 4-16 Perforated diffuser for pipelines larger than 3 ft (0.9 m) in diameter

Even the smallest leak will remove cadmium plating from chlorine tubing and fittings. The metal underneath the plating (copper, brass, or bronze) will appear reddish, and a green copper-chloride scum may appear around the edges of the area affected.

Portions of the pressure piping system (e.g., the manifold) are often painted. As a result, discoloration of the metal beneath the paint will not be apparent. To locate leaks in painted piping, look for small droplets of water that may form on the underside of joints. Small, almost invisible leaks must be located early; otherwise, the corrosion they cause will often result in a sudden and massive chlorine leak after a period of time.

The best and most reliable way to find chlorine leaks is with a chlorine detector, which is sensitive to leaks as small as 1 ppm chlorine in air. Such leaks are not usually detectable by the ammonia technique or by smell.

If a major leak requires shutdown of the system for repair, the tank valve should be closed, the yoke disconnected, and the injector left running with the auxiliary tank valve open, until any remaining chlorine gas is purged from the line. To prevent leaks, the operator should observe the following precautions:

- Install a new gasket every time a cylinder or container is changed.
- Each time a threaded fitting is opened, clean the threads with a wire brush. Then wrap them with PTFE tape or use one of the following pipe joint compounds: linseed oil and graphite, linseed oil and white lead, or litharge and glycerine. If PTFE tape is used, remove any previous remnants of tape before remaking the joint.
- Replace all chlorine supply line valves annually. Refit and repack the old ones so they are ready for use the following year.

Stiff container valves. Container valves are carefully checked before leaving the manufacturer's plant, but occasionally a valve may be stiff to turn or difficult to shut off tightly. This problem is often caused by overly tight packing. Sometimes the valve can be freed by opening and shutting it a few times. If the valve does not operate at all, set the container aside and call the supplier.

Liquid Chlorine Spill Containment and Neutralization Systems

Chlorine gas leaks can be dealt with quickly by the use of the container kits described in chapter 5. A major leak involves the release of *liquid* chlorine from its pressurized container. Liquid spill emergency systems include a confined sump for collecting the liquid and either a scrubber or an absorption tank to neutralize the liquid chlorine (Chlorine Institute 1998b). In the United States, the Uniform Fire Code (UFC) of 1988, and revised in 1991 and 1994, requires that a neutralizing system be able to handle the full contents of the largest single storage container. The UFC is not necessarily adopted in every jurisdiction and other local requirements may apply, but this regulation is usually the basis of most system designs.

A common neutralization design satisfies Section 80.303 of Article 80 of the UFC as it pertains to the indoor storage of compressed gasses (White 1999). This regulation requires the system to limit the discharge concentration of the chlorine vapor to one half of the immediately dangerous to life or health (IDLH) limit, which is 10 ppm. Usually systems are composed of the following components (Figures 4-17, 4-18):

- a chlorine storage room that is contained and connected to the neutralization equipment,
- a scrubber to provide efficient contact with the caustic solution to result in complete absorption of chorine vapors,
- a caustic tank with sufficient capacity to neutralize the largest single container,
- integrated caustic recirculation pumps and an air handling system, and
- chlorine monitoring analyzers on air discharge and in the storage room.

Sodium hydroxide (NaOH) (caustic) is used to neutralize the chlorine. The neutralization reaction is

$$\underset{\text{sodium hydroxide}}{2NaOH} + \underset{\text{chlorine gas}}{Cl_2} \rightarrow \underset{\text{sodium chloride}}{NaCl} + \underset{\text{sodium hypochlorite}}{NaOCl} + H_2O + \text{heat} \tag{4-1}$$

Therefore, for each 1 lb of chlorine to be neutralized:

1.13 lb of NaOH is required (pure chemical)
0.825 lb of NaCl is produced
1.05 lb of NaOCl is produced
0.25 lb of water is produced
temperature of NaOH will rise (amount depends on several factors)

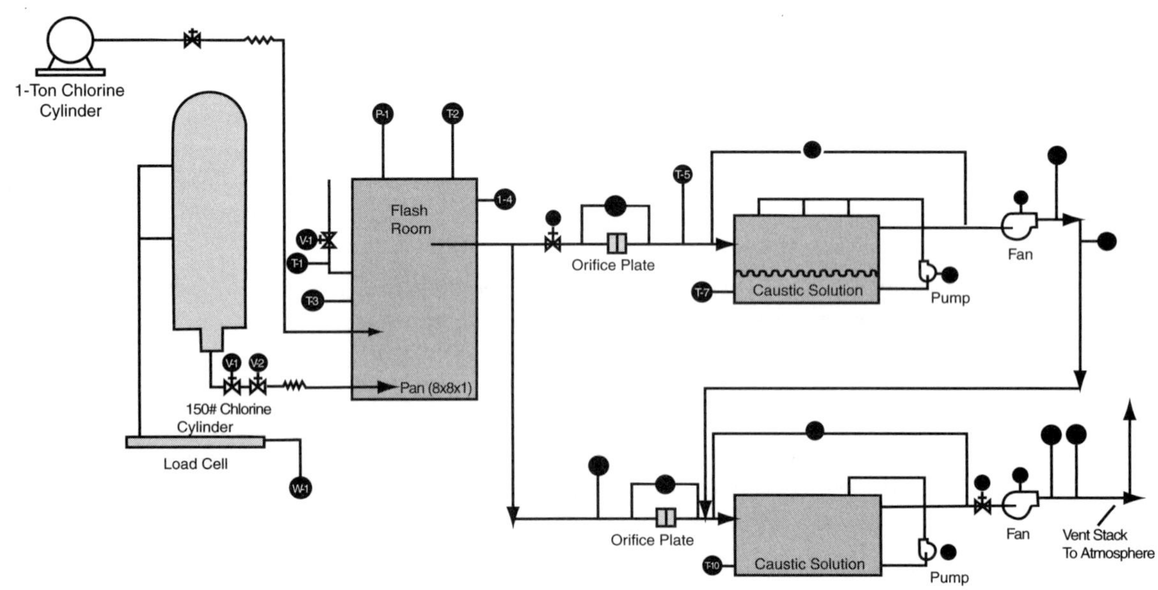

Adapted from White 1999.

Figure 4-17 Chlorine scrubber test process flow and instrumentation diagram

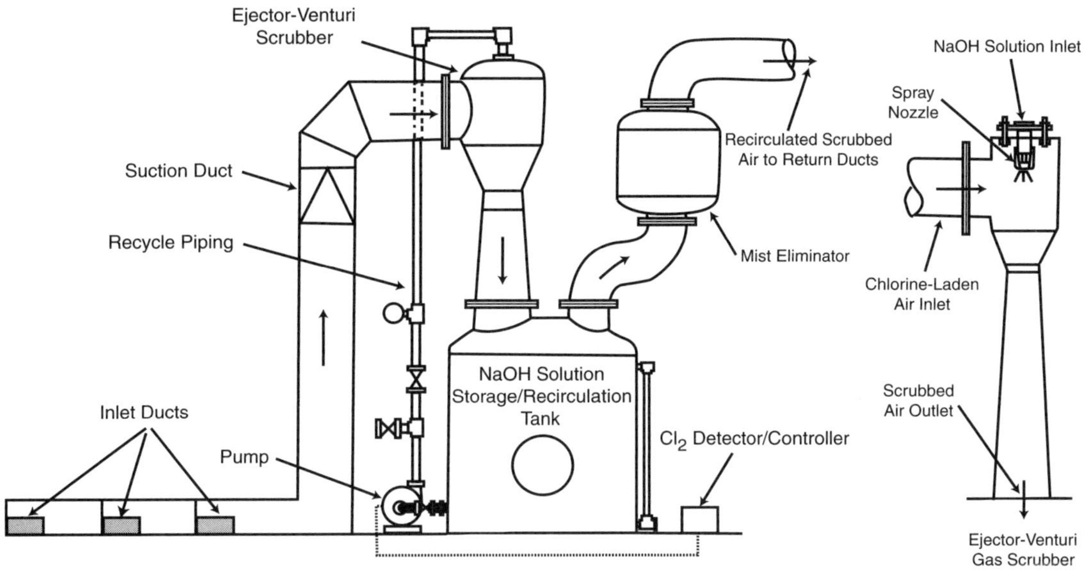

Adapted from White 1999.

Figure 4-18 Chlorine scrubber system

NaOH solutions absorb carbon dioxide from air when in storage and when testing the scrubber system. Caustic solutions are commonly delivered at 50 percent strength, but due to the low crystallization temperature of NaOH, it is usually diluted to a working solution strength of 20 percent. One consideration is to recognize that caustic solutions absorb carbon dioxide from air according to the following equation:

$$\underset{\text{carbon dioxide}}{CO_2} + \underset{\text{sodium hydroxide}}{2NaOH} \rightarrow \underset{\text{sodium carbonate}}{Na_2CO_3} + H_2O \qquad (4\text{-}2)$$

This reaction reduces the strength of the caustic solution and may make the solution inadequate to neutralize the entire contents of the chlorine container. It is important to recognize this and to test the concentration of the caustic solution periodically to verify its strength.

Examples of some of the major design considerations for a 1-ton chlorine scrubber include

- 2,100 gal NaOH,
- scrubber rated at 3,000 cfm and 550 gpm caustic flow,
- 20 percent caustic solution,
- 3,000 chlorine vapor capture rate, and
- 2,400 lb in 30 min chlorine neutralizing capacity.

Following the neutralization of a chlorine leak, the spent caustic solution (NaOCl, sodium chloride [NaCl], and NaOH) must be disposed of properly. This is often done by metering in small quantities over a period of time to a wastewater treatment plant. Another possible disposal option is to contact the caustic supplier or a professional disposal contractor for assistance. Hypochlorite that is present may be destroyed by catalytic decomposition using nickel and iron. The use of sulfites to dechlorinate the hypochlorite is not recommended because the heat from the reaction is too great due to the concentrations involved.

Courtesy of TGO Technologies, Inc.

Figure 4-19 Example of total cylinder containment

Another option to approach the possibility of a major leak from a ton cylinder is total cylinder containment. Equipment is available, and may be approved, that seals the ton cylinder in a device that is tested to contain the pressure and chlorine gas. Chlorine can still be fed in the usual way from the cylinder in case of a liquid chlorine leak. The containment vessel is removed from the site to the chlorine manufacturer for neutralization. This type of installation requires one vessel for each ton cylinder and, potentially, eliminates the need for a scrubber, as shown in Figure 4-19. There are also automatic shut off valves that can activate when chlorine is detected. These are approved for use in some areas.

LIQUID HYPOCHLORITE FACILITIES

Hypochlorite solutions are highly volatile, so the feed systems must be carefully designed or gas binding may occur. The solutions are generally delivered to the application point by gravity, chemical feed pump, or eductor system. Gravity feed systems have the advantage of simplicity. A hypochlorite storage tank provides the pressure needed and the flow is modulated using a flowmeter and a diaphragm valve. A constant-head tank is sometimes used to provide improved flow control.

Solutions of NaOCl are easily fed using chemical feed pumps (Connell 1996). If the feed rate volume is too large for chemical feed pumps, centrifugal pumps may be more economical. Chemical feed pumps with mechanically or hydraulically controlled diaphragms using solenoid, piston rod, or plunger drives that adjust stroke lengths or vary cycles or motor speed are the types most commonly used. Whatever the type, chemical feed pumps are positive displacement, volumetric pumps that increase the pump chamber volume to receive the chemical to be pumped and then decrease the pump chamber volume to push the chemical to the point of use. A series of inlet and outlet check valves control the fluid movement. In terms of operation, chemical feed pumps are typically more demanding than centrifugal pumps.

Chemical feed pumps with mechanical diaphragms, as shown in Figure 4-20, have the drive rod directly connected to the diaphragm. Although these pumps may be economical, they are also prone to diaphragm failure due to the excessive strain placed at the diaphragm/drive-rod connection point. Solenoid-operated pumps are usually mechanically connected to the diaphragm and are often used for low flows and low injection pressures.

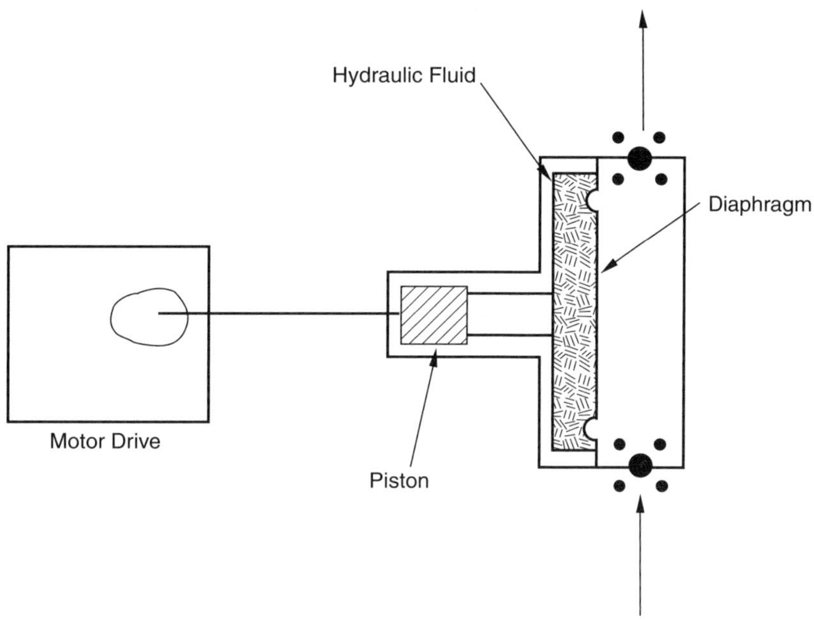

Source: Connell, 1996.

Figure 4-20 Chemical feed pump hydraulic drive

Hydraulic drive pumps overcome this action on the diaphragms by using hydraulic fluid on the drive-rod side of the diaphragm. The fluid transmits the force of the drive rod or piston uniformly across the entire area of the diaphragm and eases the strain on the diaphragm, as shown in Figure 4-20. Some pump designs use magnetic drives. Each pump type must be evaluated based on its features and the application needs before a choice is made.

Eductor systems are generally used only for larger feed rates. The eductor is used to provide a vacuum that pulls the hypochlorite from the storage tank to the injection point. The flow is controlled with a modulating valve. Compatible materials of construction are necessary to result in a reliable system.

Typical Sodium Hypochlorite Feed and Piping

The gas-producing reactions cause problems for the solution metering pump used to meter the NaOCl process. Because gas is compressible, the metering pump may become "air-bound" and eventually stop. When this occurs, the trapped gas pressure must be relieved on the discharge of the pump to allow proper operation. The piping diagram shown in Figure 4-21 depicts typical piping with vented risers to minimize any air getting trapped in the liquid end of the pump.

Precautions must be taken when designing a handling system for NaOCl to ensure the stability of the solution is maintained, including the following:

- Do not use piping systems constructed of metal (with the exception of tantalum and titanium) because metal will cause the solution to degrade due to the oxygen-producing reaction. For example, 0.5 mg/L of iron in a 15 percent solution will cause the solution to completely decompose in a few days.
- Avoid situations that would lower the pH of the solution. A pH value between 9.5 and 10.5 produces the most stable solution. Excess acidity will enable the hypochlorite to form chlorine gas and vaporize from the solution.

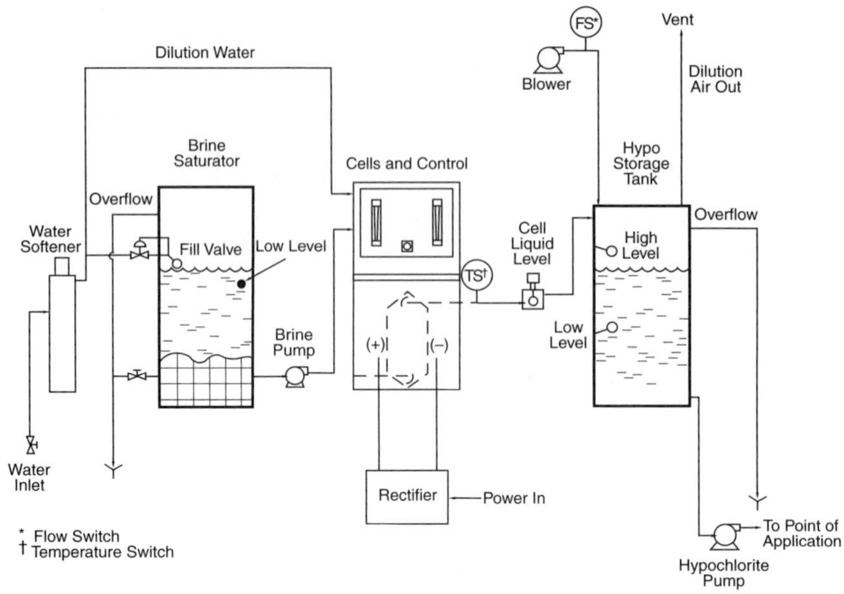

Source: Connell, 1996.

Figure 4-21 On-site hypochlorite flow diagram

- Store the solution in opaque tanks to control the storage temperature and light degradation. Ensure all parts of the storage tank are compatible with NaOCl.
- Avoid piping metering pumps with suction lift as this will increase the tendency to outgas. As the solution supply tank empties, the suction lift will increase as will the negative pressure required to move the solution into the metering pump head.
- Locate the pump as close to the supply container as possible; keep suction lines short.
- Locate the pump and the supply container as close to the point of application as practical.
- Provide a pressure relief valve in the discharge piping or purchase a pump with built-in pressure relief.
- Provide a bypass valve in the discharge piping near the pump to permit draining of the line for maintenance.
- Lay out suction and discharge piping with 90° traverses between the fixed ends and support piping vertically; allow free horizontal movement to permit pipe to expand and contract with temperature changes.
- Install unions adjacent to the pump valves in both suction and discharge lines.
- Avoid the use of braided flexible metal lines on the suction side; they can be lined with a corrugated plastic tube that can provide excellent air traps.
- Do not oversize the suction lines. Keep velocities high to sweep the entrained air/gasses along so they do not collect and cause air binding. Also, dilute the solution and pump at a higher rate to keep velocities high.

- Use a vented riser in the suction line as close to the suction valves as possible. The riser should be of a larger diameter than the suction line to ensure venting of any entrained air/gas.
- One method of reducing the effects of solution degradation is to dilute the purchased hypochlorite solution to a weaker strength. Although this will require a larger-capacity metering pump, the benefits may be worthwhile.

Operational considerations. Pump calibration is an important factor when using chemical feed pumps. The frequency of calibration and the accuracy required may dictate that a system of calibration be permanently installed in the piping system. The simplest way to calibrate a pump is to feed to a graduated cylinder using a stopwatch. Because the chemical feed pump is a volumetric feeder, the specific gravity of the chemical being used must be taken into consideration.

Also of importance is the pressure under which the system operates. Because most chemical feed pumps use small-diameter lines (½ in. [13 mm] or less), long pumping distances allow for the possibility of water hammer. High pressures at the point of application are also a concern. Piping must be appropriately designed. A solution to either of these conditions is to install an expansion chamber or pulsation dampeners in the line and review the hydraulic design to consider larger line sizes.

Maintenance and service. The area of most concern with chemical feed systems in general and hypochlorite feed systems in particular is the buildup of deposits at the inlet and outlet valves. Deposits around the valve seat prevent either valve from seating and may lead to a drop-off in feed or complete failure of feed. Valves should be inspected frequently for such buildups. Acid cleaning or water washing is usually sufficient to remove deposited materials. On each such occasion, the O-rings and gaskets used at the check valves should be replaced.

Replacement diaphragms should be readily available. The frequency of replacement will depend on operating conditions. Closing downstream valves while the pump is operating should be avoided as this may result in a high backpressure that may cause the diaphragm to fail and perhaps rupture unless a relief system is provided. The type of diaphragm used and its chemical resistance to the material being pumped is also important. The manufacturer should be consulted on the proper selection of diaphragms. Critical feed equipment should have standby pumps available. Replacement parts must be readily accessible, and staff must be trained in the maintenance of the pumps.

On-Site Hypochlorite Generation

There is increasing interest in generating NaOCl on-site due to safety considerations, simplicity of operation, and the ready availability of the raw materials required.

NaOCl generated on-site is produced in an electrolytic cell using a brine (NaCl) solution feed stock, as shown in Figure 4-22. As the solution is passed between two electrodes, a direct current is applied. The energy imparted produces Cl_2, NaOH, and hydrogen gas as a by-product (Eq 4-3). If there is a membrane between the electrodes, the chlorine gas and NaOH remain on opposite sides of the membrane. Industrially operated chlorine/caustic systems use the same principle of operation (Chlorine Institute, 2000b), only on a larger scale, to produce tons of chlorine and caustic. If there is no membrane, the two chemicals combine to form NaOCl solution, as demonstrated in Eq 2-1 in chapter 2. The concentration of the solution produced is approximately 1 percent.

$$\underset{\text{sodium chloride}}{2NaCl} + 2H_2O + \text{electric current} \rightarrow \underset{\text{chlorine gs}}{Cl_2} + \underset{\text{sodium hydroxide}}{2NaOH} + \underset{\text{hydrogen gas}}{H_2} \qquad (4\text{-}3)$$

Source: Connell, 1996.

Figure 4-22 On-site hypochlorite generation systems use only salt, water, and electric power to generate sodium hypochlorite. The systems are used at wellheads and treatment plants.

A hypochlorite solution with this low concentration does not decompose as rapidly as the industrially available concentrations of 12–13 percent used at many facilities. With low concentrations formed and the slow decomposition that results from these low concentrations, the production of sodium chlorate is considerably reduced. The raw materials used (salt, water, and electric power) are environmentally acceptable and offer little safety hazard compared to chlorine gas and NaOCl solutions.

Equipment used includes a water softener, storage tanks for the hypochlorite solutions produced, a brine saturator, an electrolytic cell, and an electrical rectifier. Some chemical feed pumps and instrumentation are also necessary.

Operating expenses depend on power, water, and salt costs. These variables must be determined locally because they are a function of local availability and are site specific. The grade of NaCl selected has a direct bearing on the quality of the hypochlorite produced and the cost. Higher grade salt may be more expensive but considerable savings may be achieved from lower maintenance costs. Each site should evaluate the cost-benefit ratio of using higher-grade salt. Capital costs depend on the type of equipment installed. For example, if locally available softened water is used, a water softener is not required. The solution produced is fed to the point of application with a chemical feed pump.

Chemical Feed Pump Control

Chemical feed pumps feeding hypochlorite solutions may be automated the same as control schemes used in gas feeders (see previous discussion). Thus, control can be manual, semiautomatic, flow proportioning, residual, or compound loop. The pump can be started and stopped manually or semiautomatically in the same fashion as the gas feeders described earlier in this chapter. The feed rate is set manually to attain the proper dosage by adjusting the stroke length, pulse, or pump speed, depending on the type available and following the manufacturer's instructions.

For automatic control with flow proportioning, residual, or compound-loop control, most modern chemical feed pumps can receive a 4- to 20-mA signal from the appropriate controller. The same controller used for gas feeders can be used to provide the correct input signal to the chemical feed pump. Most chemical feed pumps available commercially use variable DC drives that receive a 4- to 20-mA signal from the controller. The pump feed rate varies in response to the signal automatically. Most pumps adjust the feed rate by varying the pump speed, although, depending on the manufacturer, other methods may also be available. It is always critical to have the pump properly calibrated following the manufacturer's instructions and to know the strength of the solutions being fed so that the feed rate or dosage may be properly established.

Just as it is important to have the pump calibrated, it is also important to check the control elements on a regular basis and provide calibration at scheduled intervals. Such maintenance applies to the flowmeter, analyzer, pH meter, and other devices used in the disinfection control scheme. If standby equipment is on-site, regularly scheduled use of this equipment should be established and followed.

Hypochlorination Feed Problems

Two problems that commonly occur with hypochlorinators are clogged equipment and broken diaphragms.

Clogged equipment. Clogging, due to calcium carbonate ($CaCO_3$) scaling, occurs primarily in two areas of a hypochlorinator: at the pump head and in the suction and discharge hoses. Scale is most likely to form when the water used for preparing the solution has a high calcium hardness and carbonate alkalinity. Under these conditions, calcium carbonate forms and causes a scale deposit in the pump head, suction hose, and discharge hose. Scale may also form in the solution injector or diffuser.

The scale can readily be removed by pumping a dilute (5 percent) hydrochloric acid solution (also known as muriatic acid) through the pump head, hoses, and diffuser. The hypochlorite solution should be completely flushed out of the system with water before the acid is used.

An associated problem affecting the pump head is the accumulation of dissolved calcium hypochlorite (lime sludge). When the solution tank level is low, or the suction foot valve is too near the bottom of a one-tank installation, the suction hose can draw some of the undissolved chemical up into the pump head and fill the area of the head. This can result in the pump not feeding hypochlorite solution, and it can cause diaphragm rupture. To prevent these problems, it is best to use a two-tank setup when calcium hypochlorite is being used.

Broken diaphragms. The second most common problem with hypochlorinators is broken diaphragms. It is important that the operator inspect the diaphragm regularly to ensure that it is functioning properly. A visual inspection of the pump head may not reveal a broken diaphragm, but an outflow of solution from the diaphragm hose is a positive indication that the unit is functioning properly.

AMMONIA GAS (ANHYDROUS AMMONIA) FACILITIES

The principles involved in feeding ammonia gas and the equipment used for this purpose are the same as those for chlorine, with the exception of some different operating conditions and materials of construction that result from the chemical and physical characteristics of ammonia. Ammonia feeders are usually called ammoniators. Like chlorinators, ammoniators may either be vacuum-operated solution feed or pressure-operated gas feed. *Do not attempt to use chlorinators as ammoniators and do not attempt to use ammoniators as chlorinators.*

Materials of Construction

Most ammonia gas feeders are constructed of ABS with polychlorotrifluoroethylene diaphragms, but the metallic components (springs and control valves) are usually made of stainless steel. O-rings are of neoprene rubber composition.

Operational Considerations

The vacuum-operated unit creates a vacuum in the venturi, just as in the gas chlorinator, but the solution of ammonia and water formed in the venturi is highly alkaline with pH values above 12. Because waters usually contain some level of dissolved calcium and magnesium carbonates, the rapid rise in alkalinity at the venturi changes the solubility characteristics of these dissolved salts and causes them to precipitate in the nozzle and throat of the venturi. The precipitate causes a change in the hydraulic characteristics of the venturi such that the vacuum level created is altered sufficiently to either reduce or stop the feed of ammonia gas. Thus, particularly with well waters, steps must be taken to keep the feeder operational. This can be done in several ways, with varying degrees of success and with varying cost penalties. The methods normally followed for vacuum feed systems are as follows:

- A water softener is installed in the ejector water supply line. The softener reduces the calcium and magnesium contents sufficiently to provide longer venturi life before cleaning is required.
- The addition of a chelating (chemical binding) agent fed to the ejector water supply line just prior to or at the point of ammonia addition will generally produce good results. Frequently, the use of polyphosphates as the chelating agent is effective.
- Timed acid feeds have also been successfully used. The pH change that occurs due to the acid dissolves the precipitated salts and maintains the venturi in satisfactory operating condition.

The three suggested vacuum feed methods are schematically illustrated in Figure 4-23.

The formation of calcium salts creates similar operating difficulties with pressure feeders. Most pressure feeders add gas through a diffuser inserted in the water channel or clearwell. A pH change occurs at the point of addition and a white salt buildup is observed on the diffuser at the point of gas bubble formation in the water. This buildup clogs the diffuser and, in time, prevents gas feed. Steps must be taken either to prevent the buildup or to clean the system. Some methods used include the following:

- The use of a neoprene bladder fitted tightly around the diffuser. The bladder has extremely fine slits that act as valves and are tightly closed when not in use. During use, calcium and magnesium salts build up on the slits, which are the ammonia addition points, and restrict gas flow. As gas is added, the

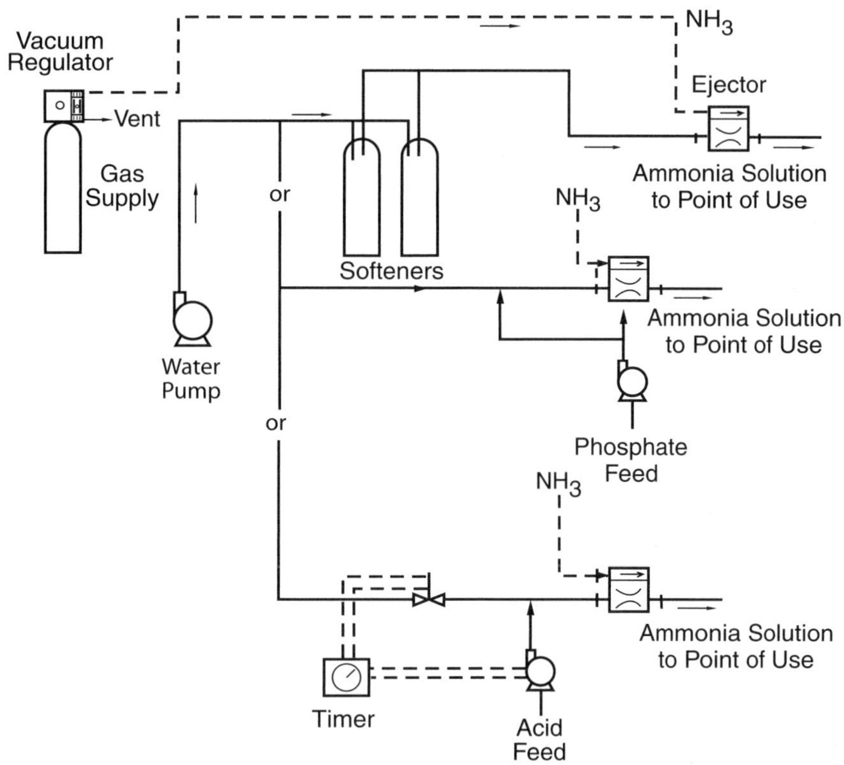

Source: Connell, 1996.

Figure 4-23 Ammonia vacuum-feed systems with antipluggage options

bladder expands, flexing the slits and breaking up the salts. As the buildup breaks up, the gas flows into the water. The bladder contracts and the cycle repeats itself.

- Occasionally, acid washes may be used. This practice is common in aeration systems that also experience the salt buildup at the aerator nozzles.

Gas pressure piping for ammonia pressure lines is usually black iron, but may be stainless steel. Flexible connectors for use in ammonia service are constructed of different materials than the traditional copper tubing used in chlorine service. Ammonia flexible connectors and valves are made of steel.

Because ammonia has a higher vapor pressure than chlorine, the liquefaction problem in gas lines that so frequently plagues the handling of chlorine may not appear as frequently with ammonia. The liquefaction preventive measures for chlorine pressure piping would generally apply to ammonia pressure piping. However, the presence of liquid ammonia in gas lines can be due to other causes. Typically, it is the result of liquid carry through from the container. It is important, then, that operating personnel be aware of the type of container provided by the chemical supplier.

Ammonia containers are provided with a dip tube connected to the container's gas-discharge valve. Operators using ammonia containers should determine the type of container and the discharge arrangement by asking the supplier for container information. Discussions with the ammonia supplier are strongly recommended to minimize the presence of liquid ammonia.

Injectors, Diffusers, and Solution Lines

Ammonia solution feed systems use injectors to draw ammonia gas from the storage cylinders and mix with dilution water. The injector should be located as close to the diffusers as possible. One problem with solution-feed ammonia systems is that the ammonia hydroxide formed when ammonia gas is combined with the solution water precipitates calcium and magnesium hardness. Therefore, the injector water should be softened to reduce the hardness to 35 mg/L or lower.

Diffusers for ammonia solution are similar to those used for chlorine. The diffuser holes can be sized to create sufficient turbulence to provide adequate mixing. Diffusers should be provided in duplicate to allow for removal for cleaning. It is also advisable to provide fittings for flushing with acid or chlorine solution to dissolve scale. Materials of construction must be resistant to ammonia solution (and acid if flushing is anticipated).

AQUA AMMONIA FACILITIES

Aqua ammonia, aqueous ammonia, and ammonium hydroxide all refer to a solution of ammonia in water. Most commercial grades of aqua ammonia are approximately 30 percent by weight ammonia. The aqueous ammonia solution contains high concentrations of hydroxide and, thus, is corrosive. Delivery, storage, and feed system components must be selected to resist this corrosive liquid.

Delivery Equipment

The US Federal Department of Transportation classifies aqua ammonia with a concentration greater than 10 percent as "corrosive." Therefore, applicable regulations for the transportation of aqua ammonia apply. Re-packagers of bulk aqua ammonia may use 55-gal steel drums, carboys, and plastic containers.

Bulk delivery may be made by cargo tanker transports having a capacity of approximately 7,000 gal. The material is transferred by pump, compressed air, or nitrogen. A pump may be part of the transport tanker or it may be part of the receiving storage system. A backflow check valve should be provided on the air or nitrogen supply line to prevent damaging ammonia vapor flow. Also, a scrubber or other vapor control system should be present to control the release of vapors from the tank. Pressure tanks should not be filled to more than 95 percent of their volume due to ammonia off-gassing. Illustrations of the most common transfer methods are shown in Figure 4-24.

Storage Facilities

Bulk storage vessels should use the applicable American Society of Mechanical Engineers (ASME) codes and requirements (ASME 1995). Tanks are equipped with relief valves as needed for the design pressure. The discharge from the relief valves shall be unobstructed to the air. Vacuum breakers are also provided and are set at 2–4 ounces of vacuum. A typical bulk storage tank is shown in Figure 4-25.

Aqua ammonia storage vessels are commonly fabricated from aluminum, stainless steel, or carbon steel. Tank linings or coatings are not commonly used for aqua ammonia storage vessels. If a coating or lining is used, it should be approved by the tank fabricator for use with aqua ammonia.

Introducing water into a storage tank containing ammonia vapors can be dangerous. The ammonia vapors are absorbed extremely rapidly causing a reduction in pressure. The vacuum-breaker supplied with the tank usually cannot allow enough air to enter to prevent a possible collapse of the pressure vessel.

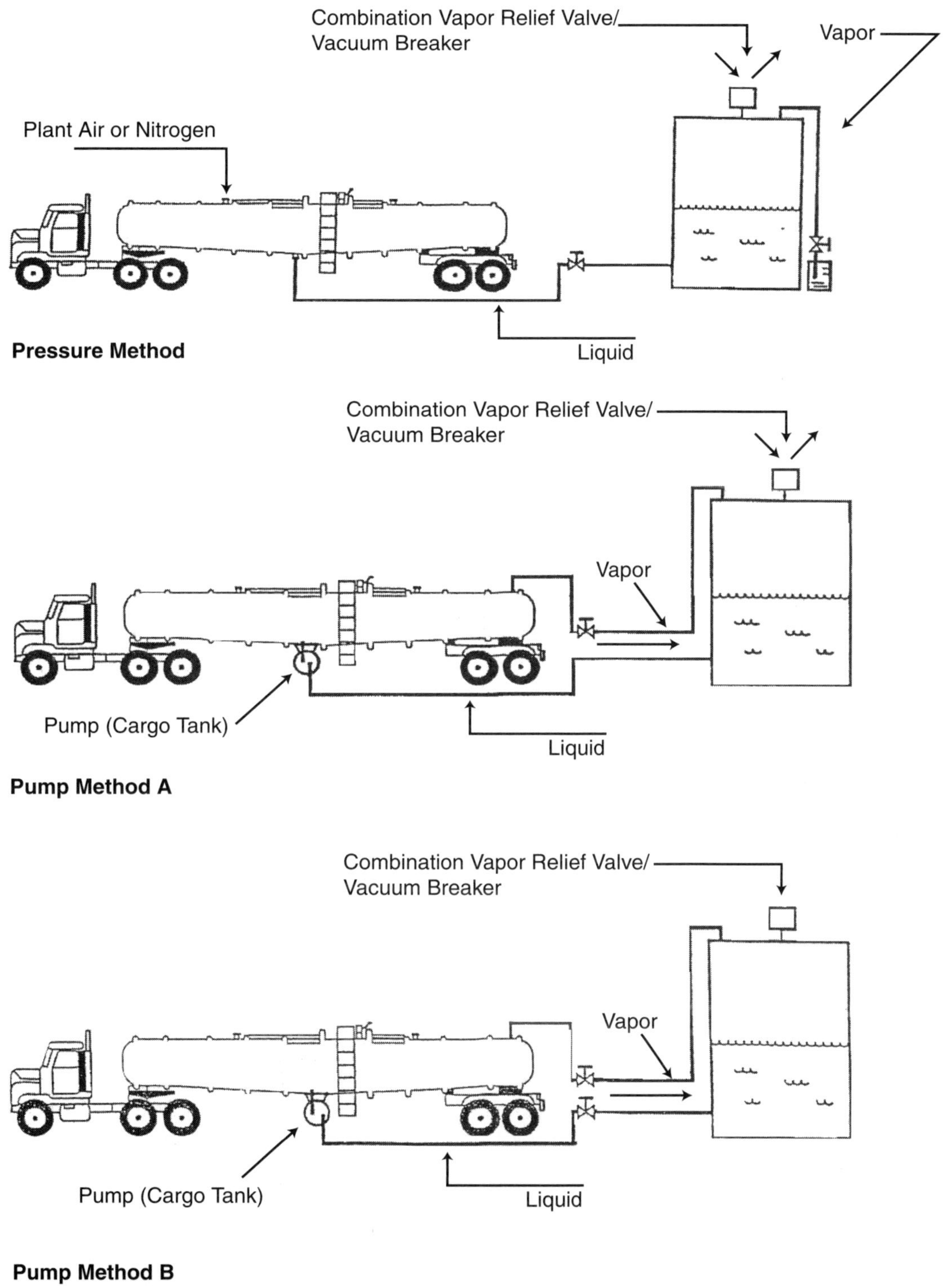

Adapted from La Roche Industries, Inc., technical bulletin.

Figure 4-24 Truck delivery methods

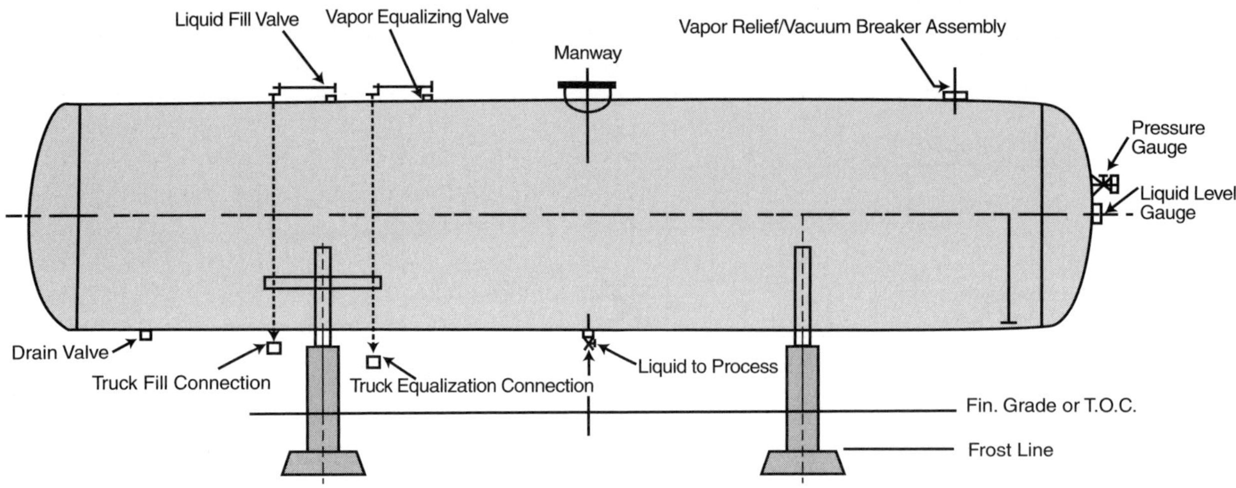

Adapted from La Roche Industries, Inc., technical bulletin.

Figure 4-25 Typical horizontal aqua ammonia storage tank

All tank openings should be equipped with shut-off valves (an exception is relief devices). Small containers should be stored in a cool place and out of the sunlight.

Storage tank maintenance usually is minimal but includes protection of external steel surfaces from corrosion and checking for leaks. Carbon steel tanks are not subject to internal corrosion after passivation. However, some rust may be present from the hydrostatic testing of the tank and from attached steel pipe and fittings. Aluminum and stainless steel tanks are not subject to internal corrosion, and painting of exterior surfaces of these tanks is optional.

COMMON FACILITIES FOR LIQUID CHEMICALS

Chemical Solution Feeders

Solutions of NaOCl, aqueous ammonia, and ammonium salts are easily fed using chemical feed pumps. If the feed rate volume is too large for chemical feed pumps, centrifugal pumps may be more economical. For descriptions of these feed systems, refer to the section, "Liquid Hypochlorite Facilities," earlier in this chapter.

Diffusers, solution lines, and mixers. The solution delivery line should be as short as possible to reduce the delay for good system control. Positive displacement chemical feed pumps used for aqua ammonia delivery allow for diffusers to be designed to provide good mixing. It is good practice to provide diffusers in duplicate to allow for their removal for periodic cleaning due to a buildup of scale from the hardness in the water. As with chlorine, ammonia solutions may need additional mixing to ensure uniform distribution and efficient reaction with chlorine where chloramination is the goal. The mixer choices and applications are similar to those discussed later in this chapter. All materials of construction of solution lines and diffusers must be resistant to corrosion from ammonia solution.

Chemical Feed Pump Control

Chemical feed pumps feeding either hypochlorite or ammonia solutions may be automated the same as control schemes used in gas feeders. Thus, control can be manual, semiautomatic, flow proportioning, residual, or compound loop. The pump can be started and stopped manually or semiautomatically in the same fashion as the gas feeders described earlier. The feed rate is set manually to attain the proper dosage by adjusting the stroke length, pulse, or pump speed, depending on the type available and following the manufacturer's instructions.

For automatic control with flow proportioning, residual, or compound-loop control, most modern chemical feed pumps may receive a 4- to 20-mA signal from the appropriate controller. The same controller that is used for gas feeders can be used to provide the correct input signal to the chemical feed pump. Most commercially available chemical feed pumps use variable DC drives that receive a 4- to 20-mA signal from the controller. The pump feed rate varies in response to the signal automatically. Most commercially available pumps adjust the feed rate by varying the pump speed, although, depending on the manufacturer, there may be other methods. It is always critical to have the pump properly calibrated following the manufacturer's instructions and to know the strength of the solutions being fed so that the feed rate or dosage can be properly established.

Just as it is important to have the pump calibrated, it is also important to check the control elements on a regular basis and provide calibration at scheduled intervals. Such maintenance applies to the flowmeter, analyzer, pH meter, and other devices used in the disinfection control scheme. If standby equipment is on-site, regularly scheduled use of this equipment should be established and followed.

Ammonia dosage control considerations. Ammonia feed systems generally use only one control scheme. In the chloramination process, the control of the chlorine-ammonia ratio at 4.5:1 to 5:1 is of paramount importance, because the ammonia feeder is tied directly to the chlorine residual and the water flow. Thus, the compound-loop control would be the most logical choice for the feeder. Some facilities, however, may use flow-proportioning control and manually set the chlorine-ammonia ratio. An example of this type of application is a well-water chlorination system in which ammonia is added to well water immediately after chlorination. For this application, manual ammoniation control with a semiautomatic on/off system is appropriate.

The nature of ammonia feed systems is such that a more problem-free performance may be attainable by using a pressure feed system. Although using pressure feed for ammonia sacrifices the safety of the all-vacuum system, the relative safety of handling ammonia, compared to chlorine, combined with the service-free system offered by ammonia pressure feed, more than balances the loss of all vacuum safety of feeding ammonia.

Therefore, ammonia feed systems will use automatic control schemes and may use either manual or semiautomatic systems. Ammonia monitors may also be used to provide a signal with flow input for compound-loop control. All systems require the maintenance of the proper chlorine-ammonia ratio.

ANCILLARY EQUIPMENT

Many pieces of equipment are used in conjunction with chlorine gas feeders and hypochlorite solution feeders. Only equipment with control, residual measurement, or safety considerations are discussed here.

Chlorine Residual Analyzers

Of all the ancillary items used in water treatment systems, the most important is the chlorine residual analyzer. The enactment of the Surface Water Treatment Rule (SWTR) dictates the continuous measurement of chlorine residual in the water. Depending on the size of the utility, periodic sampling may be satisfactory. However, most prudent water utilities are following a course of continuous measurement and are considering control of chlorine feed to meet a preset value. The function of the analyzer is to measure the chlorine residual in the water so that an indication of the

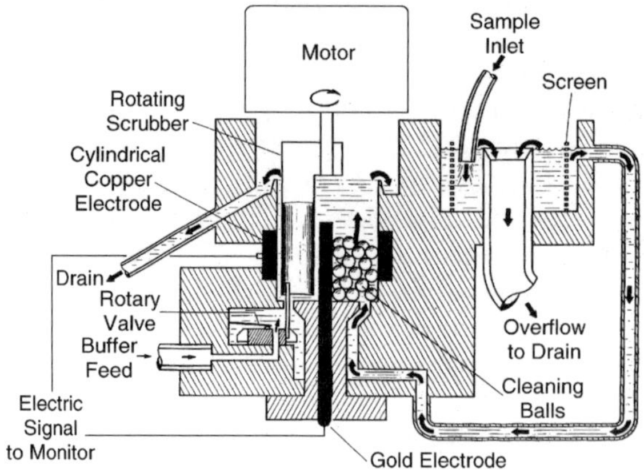

Courtesy of Severn Trent Services.

Figure 4-26 Automatic chlorine residual analyzer

residual is available on a continuous basis and a signal can be sent to a controller to pace a chlorinator. Generally, the most commonly used analyzers are one of two types: amperometric or colorimetric.

An amperometric analyzer, shown in Figure 4-26, typically consists of two electrodes that are immersed in a continuous water sample. The electrodes are made of two dissimilar metals, e.g., gold and copper, that measure a change in current flow between them that is directly proportional to the amount of chlorine residual in the water stream. While most units use the electrochemical potential between the two metals as the driving force, some may have low voltage impressed across the electrodes.

An amperometric analyzer may be used to measure either total or free residual chlorine. For free chlorine, a buffer is added to maintain a constant pH of approximately 4 to enhance signal stability. An iodide salt and buffer are added to permit the measurement of total chlorine residual. The iodide reacts with both the free and combined chlorine residuals to permit measurement of both species. The electrical signal produced at the electrodes is converted to a 4- to 20-mA analog signal for local or remote indication and recording or fed to a controller to be used for residual or compound-loop control.

Colorimetric analyzers consist of a photoelectric cell and a light source that detects a variation in color produced in the sample stream on addition of a reagent specific for chlorine, free or combined. The reagent will react with the residual chlorine to produce a color whose intensity will vary with the amount of chlorine present. The color change is measured in the cell and a conditioned signal is sent to an indicator, recorder, or controller. Most colorimetric analyzers, although indicating continuously, sample on a timed basis and the indicated value of chlorine is changed on a timed-cycle basis. The cycle depends on the reaction time of the chemical and the time to reach a stable color.

A recently developed instrument uses a probe that can be positioned permanently in the water to be sampled or held in a stilling well to which the sample is pumped. These units contain electrodes suspended in an electrolyte and separated from the sample by a membrane. As chlorine molecules pass through the membrane, the amount of current produced varies directly with the chlorine present.

As with any analytical instrument, calibration, service, and maintenance are of primary importance. Chemicals must be maintained in a fresh and ready state. Cleanliness of the measuring cell is critical. Most amperometric devices use in situ cleaning devices, such as small plastic balls permanently located in the cell or periodically added abrasives that are pushed against the electrode surfaces and agitated to maintain a constant state of cleanliness. A colorimetric device generally requires flushing with water or cleansing fluid and periodic cleaning of the photoelectric cell's outside surface.

Periodic calibration of all types of chlorine residual analyzers is recommended, and it is a good practice to do this weekly. (Many plants routinely check calibration daily.) There may be occasions when a situation at a site would dictate a more frequent calibration. Signal drift may be an indication that the measuring cells require cleaning. A calibration check after each cleaning is also recommended. At a minimum, the recommendations of the manufacturer should be followed.

In all cases, the cleanliness of the sample is of utmost importance. Filters or flushing Y-strainers will aid in keeping the cell clean. These same filters and strainers will also require servicing to ensure the sample is not becoming contaminated.

Gas Detectors

Devices that measure a gas in the surrounding atmosphere are called *gas detectors*. State and local safety regulations and some building codes may require facilities to install gas detectors that are sensitive at levels of gas concentrations in the range of the maximum allowable concentrations (MAC). All storage and use areas should have active gas detection systems in operation. It should be remembered that a detector will not locate the leak but will simply tell the user that gas is present. It is up to the user to respond to the alarm condition and to locate and determine the nature of the leak. The recommended sensitivity levels for chlorine and ammonia gas detectors are both 1 ppm/volume.

The great majority of detectors are of the electrochemical type and are capable of detecting small concentrations of gas in the surrounding air by a change in the conductivity of probes that are immersed in an electrolyte. The electrolyte will absorb the chlorine or ammonia and indicate the change as an electrical signal.

Effective use of the signal is important. Activation of a local external or control room signaling device (e.g., a horn or light) is recommended, and a connection to the local emergency planning operation or hazardous material (Hazmat) organization may be required by local regulations. Local fire departments usually respond to these emergencies. Most local fire codes require that an emergency plan be submitted, approved, and coordinated with the local group. Some fire codes may require the use of exhaust scrubbing systems, but organizations such as the Chlorine Institute recommend a risk analysis be conducted to determine the need.

A risk analysis considers such factors as quantity in use; proximity to other facilities, residences, and factories; prevailing winds; availability of trained personnel; and preparation of emergency plans to evaluate the need for containment or treatment systems.

Most current code requirements permit the storage and use of up to four 150-lb (68-kg) cylinders in a control area if the rooms are equipped with sprinklers and constructed with materials suitable for a 1-hr fire wall rating. When meeting this requirement, the facility may operate without the need of a containment system (scrubber). In locales covered under the standard or national Building Officials and Code Administrators of America (BOCA) codes, automatic shutoff devices, emergency kits, containment vessels (coffins), or other containment methods may be employed in lieu of scrubbers. It is important to determine what the local code requirements are relative to these matters.

Automatic Changeover Devices

Most state regulations require 100 percent equipment standby for chlorination facilities. In addition, the ability to automatically change from an empty supply source to a full standby supply is often required. The use of automatic changeover devices provides the treatment plant operator with some assurance that water will not be sent to the distribution system without a chlorine residual.

Automatic changeover systems, illustrated in Figures 4-27 and 4-28, are available to either mechanically or electrically change from one source to another. Many installations require the ability to switch gas feed sources. The majority of mechanical devices are vacuum operated and use the high-vacuum level or vacuum difference developed when the gas supply is depleted to switch from the empty, on-line source to the full, standby source. These sources may either be cylinders or ton containers. Larger installations that use liquid chlorine from ton containers may also desire to switch from one liquid supply source to another, as shown in Figure 4-29. The use of pressure sensors and suitable electrical interlocks provide the necessary capability. Switching from tank car to tank car can be done but is limited due to US Department of Transportation (DOT) regulations requiring manned attendance not only during operation but also during switching. This usually defeats the purpose of automatic changeover. Manual changing of tank cars is the industry recommendation as the safest procedure.

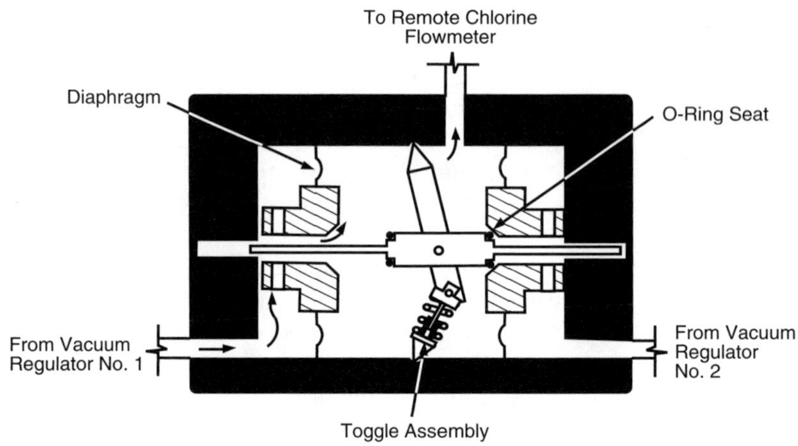

Courtesy of Severn Trent Services.

Figure 4-27 Automatic switchover unit

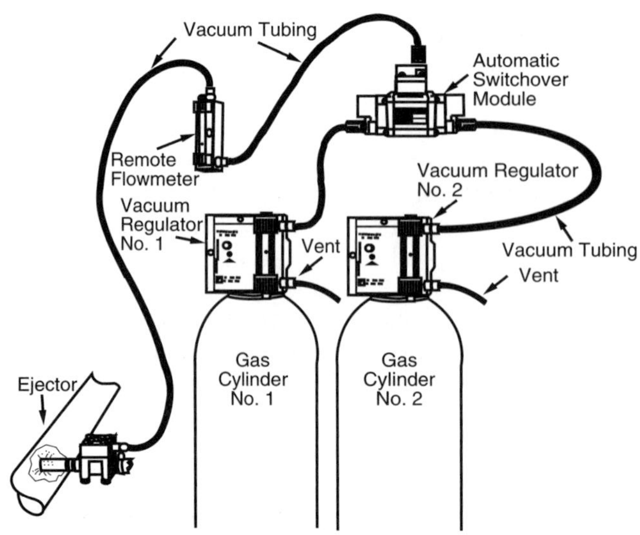

Courtesy of Severn Trent Services.

Figure 4-28 Typical installation of switchover system

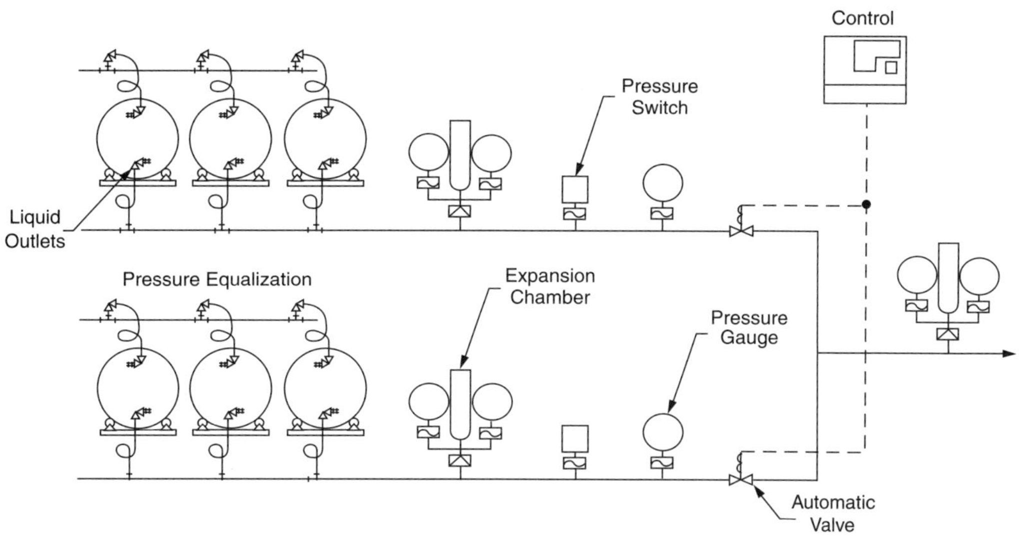

Source: Connell, 1996.

Figure 4-29 Liquid chlorine changeover system

Chlorine and Ammonia Vaporizers

A vaporizer, shown in Figure 4-30, uses an external heat source to convert liquid chlorine or ammonia to gas at a desired rate. Vaporizers are used at large installations that require gas in quantities greater than that capable of being produced by the supply without an external heat source (Chlorine Institute 1997). In general, when the chlorine feed rate exceeds 2,000 lb/d (40 kg/hr), or chlorine is only available in tank cars, trucks, or on-site storage, and when the manifolding of multiple sources in a sufficient number to permit gas withdrawal is not possible nor practical, chlorine vaporizers must be used. When the desired feed rate of ammonia exceeds 1,000 lb/d (20 kg/hr), or the ability of the source to provide sufficient heat to meet the gas flow rate desired cannot be met, ammonia vaporizers must be used.

The operation of chlorine or ammonia vaporizers does not increase the pressure of the chemical in either the liquid or gas portion of the system. However, because vaporizers operate under pressure, all vaporizers must be designed to meet the ASME code (ASME 1995) for pressure vessels. In addition, because ammonia is classified as a class 1, group D, division 1 gas, there are National Electrical Code requirements specific for ammonia. Thus, additional evaluation of the installation and provisions in the equipment design are required.

The vaporization chambers in chlorine vaporizers are usually immersed in a water bath with heat supplied by electric heaters. Some types have an external heater and a pump to circulate the heated water from the heater to the vaporizing chamber. Alternatively, low-pressure steam vaporizers may also be used, although their presence in the water industry is not as common as the electrically heated type. A typical electrically heated, water bath vaporizer is illustrated in Figure 4-31.

Vaporizers consist of a pressure vessel, containing liquid chlorine or ammonia, surrounded by water. The liquid chlorine or ammonia absorbs heat through the pressure vessel walls until the liquid is sufficiently heated to vaporize and exit the chamber. There is neither a chlorine nor ammonia liquid level control because the vaporizer level is maintained by the feed rate established by the downstream gas feeding device. Liquid carryover can occur whenever the feed rate exceeds the ability of the vaporizer to provide sufficient heat or there is a buildup of dirt left in the chamber over periods of vaporization.

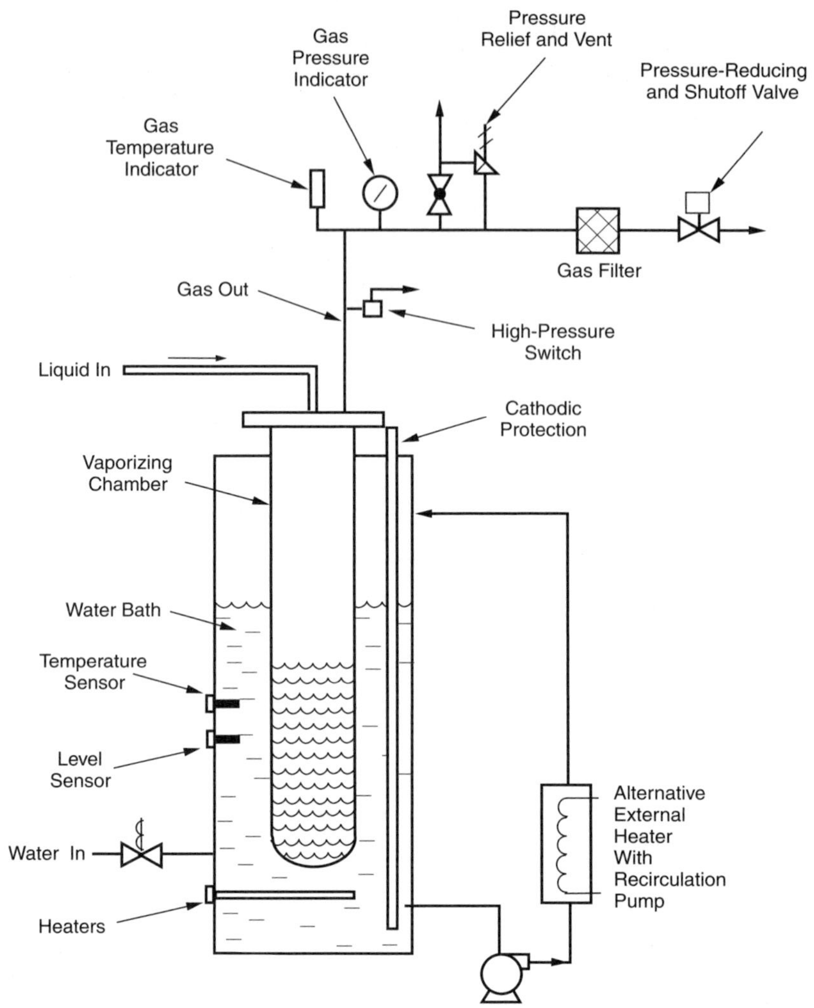

Source: Connell, 1996.

Figure 4-30 Chlorine/ammonia vaporizer

The vaporizer must be equipped with several devices for safe operation, including the following:

- **Pressure-reducing and shutoff valve (PRV).** This valve will close on low water temperature, low water level, or loss of electric power. The PRV also functions as a device to prevent liquid carryover because the drop in pressure across the valve provides superheat and protects the downstream feeder from liquid chlorine.
- **Pressure-relief valve.** The ASME Pressure Vessel Code requires a pressure-relief device to protect against over-pressurization of the vaporization system.
- **Temperature control.** Both the water bath and the piping for gaseous chlorine or ammonia will have sensors to determine the respective temperatures for control or monitoring purposes. In the case of the water bath, the water temperature is controlled in a narrow range sufficient to vaporize the chlorine. High water temperature will shut off the power to the water heaters and shut the system down. The temperature of the gas exiting the chamber must be

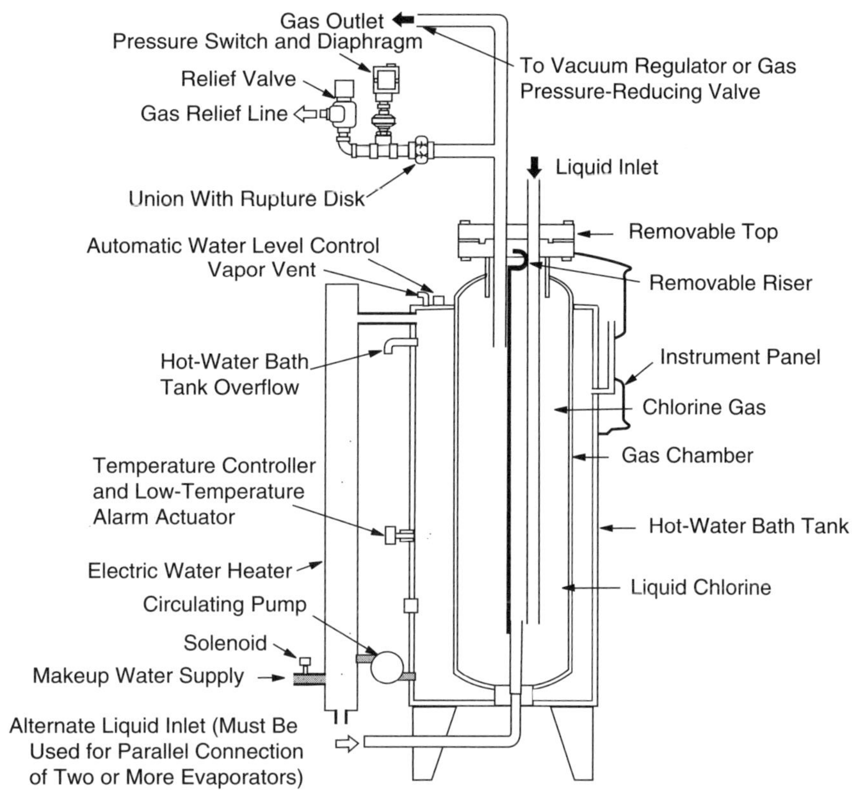

Courtesy of US Filter/Wallace & Tiernan.

Figure 4-31 Chlorine vaporizer

monitored to ensure sufficient superheat is obtained to minimize the possibility of downstream reliquefication and to provide an indication for scheduling chamber cleanout. Comparing the off-gas temperature from the vaporizer to its position on the chlorine or ammonia vapor pressure curves can be used to identify sludge buildup in the vaporizing chamber; 10–20°F (5–10°C) of superheat is considered sufficient. Less than that is an indication of possible sludge buildup or heater failure.

- **Water-level switch.** A sufficient level of water must be maintained so that the heat generated by the electrical heaters is transferred to the water bath and then to the liquid chlorine for vaporization. The presence of the water surrounding the chamber acts to ensure that the temperature of the chamber and chlorine are maintained below 180°F (82.2°C). A low water level in the water bath will shut the system down and turn off electrical power.
- **Cathodic protection.** Protection against corrosion of the vaporizing chamber and the water bath tank is provided on the water side by donor electrodes. An indicating ammeter is used in an electrical circuit to monitor protection and the presence of the electrodes.
- **High-pressure switch.** In the event of high pressure, the high-pressure switch disconnects the vaporizer heater power supply and provides an alarm contact to permit an external alarm signal.

The Chlorine Institute may provide written information on the satisfactory design, proper installation, safe operation, and recommended minimum maintenance of vaporizers (Chlorine Institute 1997).

Expansion Chambers

All liquid chlorine lines must be provided with expansion chambers to protect the liquid line from over-pressurization. An expansion chamber must be installed in a liquid line between any two valves. The chamber must be sized to protect up to 20 percent of the line volume. All expansion chambers are equipped with rupture disks that relieve into the empty chamber. A pressure gauge and pressure switch are located between the rupture disk and the expansion chamber to provide a visual indication of the pressure in the chamber and a contact closure that would activate an alarm signal to alert operating personnel of a potentially unsafe situation. The Chlorine Institute's Drawing 183 dealing with expansion chambers should be reviewed (Chlorine Institute 1994).

Mixers and Mixing

In some applications, it may be necessary to provide additional mixing at the point of application to ensure rapid and complete chlorination/ammoniation. Many types of mechanical and static mixers are suitable for this purpose. Figure 4-32 illustrates the use of a propeller type of mixer in an open channel in which chlorine solution is distributed with a perforated pipe diffuser. Static mixers may be used to advantage where only a short distance is available for mixing before a sampling location. Several types of jet mixers have been constructed for use with chlorine/ammonia solutions. These devices generally require the use of higher pressure, usually provided by an auxiliary pump, to force the solution through a jet nozzle assembly. Regardless of which device is selected, the materials of construction of all mixing components must be resistant to chlorine/ammonia solutions.

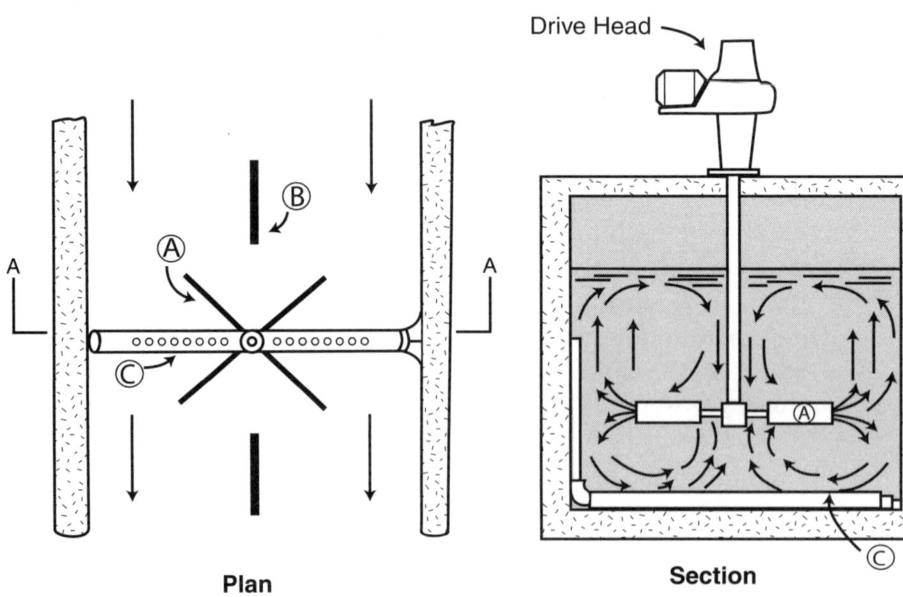

Adapted from White 1999.

Figure 4-32 Propeller mixer in open channel flow showing location of baffles. A = radial flow impeller; B = baffles; and C = chlorine diffuser.

REFERENCES

ASME (American Society of Mechanical Engineers). 1995. *Boiler and Pressure Vessel Codes.* Section VIII, Rules for Construction of Pressure Vessels. New York: ASME.

Chlorine Institute. 1994. Manifolding Ton Containers for Liquid Chlorine Withdrawal, Drawing DWG 183. Edition No. 3. Arlington, Va.: The Chlorine Institute.

———. 1997. *Chlorine Vaporizing Systems,* 5th ed. Arlington, Va.: The Chlorine Institute.

———. 1998a. *Piping Systems for Dry Chlorine,* 14th ed. Arlington, Va.: The Chlorine Institute.

———. 1998b. *Chlorine Scrubbing Systems,* 2nd ed. Arlington, Va.: The Chlorine Institute.

———. 1999. *Water and Wastewater Operators Chlorine Handbook,* 1st ed. Arlington, Va.: The Chlorine Institute.

———. 2000a. *The Chlorine Manual,* 6th ed. Arlington, Va.: The Chlorine Institute.

———. 2000b. *The Sodium Hypochlorite Manual,* 2nd ed. Arlington, Va.: The Chlorine Institute.

Connell, G.F. 1996. *The Chlorination/Chloramination Handbook.* Denver, Colo.: American Water Works Association.

Doolittle, J.S., and A.H. Zerban. 1955. *Engineering Thermodynamics.* 2nd ed. New York: McGraw Hill.

Von Huben, H. 1995. *Principles and Practices of Water Supply Operations: Water Treatment*, 2nd ed. Denver, Colo.: American Water Works Association.

White, G.C. 1999. *Handbook of Chlorination and Alternative Disinfectants.* New York: John Wiley & Sons.

This page intentionally blank.

Chapter 5

Chlorine and Ammonia Safety

SAFETY-RELATED CHEMICAL AND PHYSICAL PROPERTIES—CHLORINE

A complete discussion of the chemical and physical properties of chlorine is provided in chapter 2. The information discussed in this chapter is of those properties that are directly related to the safe handling of chlorine.

Liquid/Gas Volume Relationship

One volume of liquid chlorine yields approximately 460 volumes of chlorine gas. For example, 1 pound (or approximately 11 fluid ounces) of liquid chlorine yields approximately 5.4 cubic feet of 100 percent chlorine gas (Cl_2) when vaporized at a normal temperature (70°F [21.1°C]) and atmospheric pressure. Therefore, one 150-lb cylinder would completely fill a 10 × 10 × 8-ft room with 100 percent chlorine gas.

Liquid/gas temperature effect. The vaporization of liquid chlorine on skin or clothing may reduce the temperature enough to cause frostbite (even through high-quality protective clothing), cause the fogging of protective face masks, or the freezing of footgear to the ground.

Reaction with water. Chlorine is only slightly soluble in water, in which it forms a weak solution of hydrochloric acid (HCl) and hypochlorous acid (HOCl). (Refer to Eq 3-1 in chapter 3.) Chlorine hydrate, a greenish ice-like substance ($Cl_2 \cdot 8H_2O$), may form as crystals below 49.3°F (9.6°C) at atmospheric pressure. It can also form at higher temperatures if the chlorine is at an increased pressure. These crystals can interfere with the proper operation of chlorination systems.

Reactions with metals. Below 250°F (121°C), iron, copper, steel, lead, nickel, platinum, silver, and tantalum are resistant to dry chlorine (gas or liquid state). At ordinary temperatures, dry chlorine reacts (often violently) with aluminum, arsenic, gold, mercury, selenium, tellurium, tin, and titanium. Carbon steel ignites at 483°F (251°C) in a chlorine atmosphere.

Wet chlorine forms acids and is very corrosive to most common metals. Platinum, silver, and tantalum are resistant to both wet and dry chlorine. Titanium is unique because it is resistant to wet chlorine but cannot be used in contact with dry chlorine. Experts should be consulted when dealing with systems using wet chlorine.

Reactions with ammonia/ammonia compounds. Chlorine should be segregated from ammonia and ammonia compounds because potentially explosive or violent reactions could result in the event of a chlorine release.

Reactions with organic chemicals. Chlorine reacts with many organic compounds. Some of these reactions can be violent or explosive, including those with oils, greases, solvents, and other hydrocarbons. The separation of these materials during storage and use is essential to safety. This is especially important when new components including piping are added to the chlorine system. Even thin layers of oils and greases can react violently.

Flammability. Chlorine, gas or liquid, is nonexplosive and nonflammable; however, like oxygen, it is an oxidizer and is capable of supporting combustion of certain substances. Many organic chemicals react readily with chlorine, sometimes violently.

Physiological Effects of Chlorine Gas Exposure

Chlorine is a potential irritant to the eyes, skin, mucous membranes, and the respiratory system. The impact of exposure to chlorine is both concentration and time dependent. As the duration of exposure or the concentration increases, the affected individual may become apprehensive and restless, with coughing accompanied by throat irritation, sneezing, and excess salivation. At higher levels, vomiting associated with labored breathing may occur. In extreme cases, difficulty in breathing may progress to the point of death through suffocation. An exposed individual with a preexisting medical or cardiovascular condition may have an exaggerated response. Anyone exhibiting these symptoms should see a qualified healthcare provider immediately as his or her condition may deteriorate over the next few hours. Proper personal protective equipment (Chlorine Institute 2001) must be utilized to protect against the harmful effects of chlorine exposure.

Exposure level guidelines and definitions. The Occupational Safety and Health Administration (OSHA) has established permissible exposure limits (PELs) (CFR 1910.1000–1910.1052) for regulating employee exposure to numerous chemicals. Similarly, the American Conference of Governmental Industrial Hygienists (ACGIH 2001) has established threshold limit value (TLV) guidelines (Chlorine Institute 1996) and the National Institute for Occupational Safety and Health (NIOSH) has developed recommended exposure limits (RELs) (CGA 1984) (Table 5-1). These may be expressed as time-weighted averages (TWAs), short-term exposure limits (STELs), ceilings, or a combination of these:

- TWA is the employee's average airborne exposure in any 8-hr shift in a 40-hr work week that should not be exceeded.
- STEL is the employee's 15-min time-weighted average exposure that should not be exceeded at any time during a work day. In some cases, an STEL of another time limit, e.g., STEL (30), may be specified.
- Ceiling is the employee's exposure, which shall not be exceeded during any part of the work day. If instantaneous monitoring is not feasible, the ceiling shall be assessed as a 15-min TWA, which shall not be exceeded at any time during a work day.
- Immediately dangerous to life or health (IDLH) represents the maximum concentration from which, in the event of respiratory equipment failure, one could

Table 5-1 Chlorine exposure levels and effects on humans

Exposure Level (ppm*)	Effect
0.02–0.2	Odor threshold (varies with individuals)
< 0.5	No known acute or chronic effect
0.5	TLV–TWA REL–Ceiling
1	PEL–Ceiling TLV–STEL ERPG-1
1–10	Irritation of the eyes and mucous membranes of the upper respiratory track. Severity of the symptoms dependent on the concentration and length of the exposure.
3	ERPG-2
10	IDLH
20	ERPG-3

Source: Chlorine Institute.

*All ppm are measured in vol/vol.

escape within 30 min without a respirator and without experiencing any escape-impairing (e.g., severe eye irritation) or irreversible health effects.

The American Industrial Hygiene Association has developed emergency response planning guideline (ERPG) values that are intended to provide estimates of concentration ranges in which one reasonably might anticipate observing adverse effects as described in the definitions for ERPG-1, ERPG-2, and ERPG-3 as a consequence of exposure to the specific substance. The definitions of ERPG-1, ERPG-2, and ERPG-3 are as follows:

- **ERPG-1:** The maximum airborne concentration below which it is believed nearly all individuals could be exposed for up to 1 hr without experiencing more than mild transient adverse health effects or without perceiving a clearly defined, objectionable odor.
- **ERPG-2:** The maximum airborne concentration below which it is believed nearly all individuals could be exposed for up to 1 hr without experiencing or developing irreversible or other serious health effects or symptoms that could impair an individual's ability to take protective action.
- **ERPG-3:** The maximum airborne concentration below which it is believed nearly all individuals could be exposed for up to 1 hr without experiencing or developing life-threatening health effects.

Nonrespiratory effects of exposure to gaseous chlorine. Gaseous chlorine absorbs in water to form both HCl and HOCl. Chlorine gas can dissolve in body moisture (i.e., perspiration) to form these acids. At 3,500 ppm (mg/L) chlorine in air, the pH of moisture on the skin would be approximately 4. A pH of 4 is comparable to carbonated water. While a burning sensation and skin irritation can occur due to such exposure, a review of the literature has provided no specific human data to determine the concentration of chlorine required to produce such effects. Irritation of the eyes, when exposed to gaseous chlorine, begins to occur at the 1- to 10-ppm level.

Nonrespiratory effects of exposure to liquid chlorine. Liquid chlorine is a liquefied compressed gas. At atmospheric pressure, liquid chlorine vaporizes at –34°C (–29°F). Typically, chlorine is stored in vessels as a liquid at atmospheric or elevated pressures. Liquid chlorine will cause eye and skin burns upon contact, similar to frostbite.

Chlorine Odor

Chlorine gas has a greenish-yellow color. It has a characteristic disagreeable and pungent odor, similar to chlorine-based laundry bleaches, and is detectable by smell at concentrations as low as 0.02 to 0.2 ppm. The odor of chlorine is unpleasant and serves to both warn those exposed to even relatively low concentrations and make them want to vacate the area. These features of chlorine are helpful in preventing serious exposures, but they should not be used instead of chlorine detectors and alarms placed in appropriate areas.

Personal Protective Equipment

Personal protective equipment (PPE) recommendations assume that the facility has not performed a detailed job safety analysis (JSA) of the specific task being performed. If such a detailed JSA of the specific task has been performed and documented, and it concludes that a different level of PPE will protect the employee(s) performing the work, such different levels of PPE are compatible with the purposes and intent of these recommendations. In all cases follow the requirements of regulations that apply at your facility.

This section covers the recommended PPE for performing the specified tasks involving chlorine liquid or gas below 120°F (49°C). (For additional information, see Table 5-3 later in this chapter.) Because all chlorine containers have both a liquid and gas phase, we recommend that PPE be considered for protection against chlorine liquid.

Employees with respiratory diseases or reduced respiratory capacity should avoid working in situations in which chlorine exposure is possible. Chlorine users should adopt a medical surveillance program suitable to their needs.

In the areas where chlorine is stored or used, no specialized clothing is required for workers performing routine plant operations. However, long pants, shirts with sleeves, safety glasses with side shields or goggles approved for use with hazardous chemicals, hard hats, and safety shoes should be worn or be available as dictated by plant practice and procedures.

Generally, all personnel entering areas where chlorine is stored or handled should carry or have immediately available an escape-type respirator. Chemical cartridge or full-face canister gas masks offer adequate temporary protection provided the oxygen content in the air is greater than 19.5 percent and the chlorine concentration does not exceed the rated capacity of the respirator. The need to protect the eyes from chlorine should be part of the evaluation of appropriate respiratory equipment, because some types of respirators also protect the eyes and additional protection is not needed if these are used.

A pressure-demand self-contained breathing apparatus (SCBA) with full-face piece, shown in Figure 5-1, is required for performing tasks when chlorine may be present unless air sampling verifies the chlorine concentration is such that a lower level of respiratory protection is adequate. Emergency responders must have regularly scheduled and documented training to ensure competency with SCBA. The SCBA apparatus should be located on-site or at an acceptable location (but not in an area likely to be exposed to chlorine gas or liquid). If arrangements have been made to use an approved outside emergency response group, then the responders and apparatus may be located off site. Fit testing and regular maintenance programs for respirator equipment are required and must be documented (29 CFR).

Specific Recommendations

This section addresses only the need for PPE in connection with initial line breaks, material sampling, and the routine operations of unloading containers performed by

Figure 5-1 Air pack with positive-pressure mask

treatment plant personnel. Emergency response operations are covered later in this chapter.

Initial line break. An initial line break is defined as the first-time opening of a line or section of a line, vessel, or other equipment that contains or previously contained chlorine. When a system returned to chlorine service is opened to the atmosphere again, that event is also considered an initial line break. An initial line break is considered a maintenance activity and does not include connecting or disconnecting containers for loading or unloading purposes or material sampling activities.

Recommendations

If the specific initial line break currently being performed has been carried out periodically in the past and it has been demonstrated that the evacuation techniques and the maintenance procedures utilized will result in chlorine concentrations no more than the capability of the respirator, follow the recommendations outlined here

- Chlorine gas Use a full-face air-purifying respirator approved for protection against chlorine.
- Chlorine liquid Use a full-face air-purifying respirator approved for protection against chlorine. Use gloves for thermal (cold) protection.

If the previous criteria have not been met:

- Chlorine gas Use SCBA or a full-face air supply respirator with an auxiliary self-contained air supply (escape air provision).
- Chlorine liquid Follow Enhanced Level B protection procedures. (More information about Enhanced Level B is provided later in this chapter.)

Material sampling. If the specific sampling task has been periodically undertaken in the past and industrial hygiene sampling results demonstrated that the techniques being utilized may result in chlorine concentrations no more than the capability of the respirator, follow the recommendations outlined here.

Recommendations

- Chlorine gas — Use a full-face air-purifying respirator approved for protection against chlorine.
- Chlorine liquid — Use a full-face air-purifying respirator approved for protection against chlorine. Use gloves for thermal (cold) protection.

If the previous criteria have not been met:

- Chlorine gas — Use SCBA or a full-face air supply respirator with an auxiliary self-contained air supply (escape air provision).
- Chlorine liquid — Use SCBA or a full-face air supply respirator with an auxiliary self-contained air supply (escape air provision). Use gloves for thermal (cold) protection (Figure 5-1).

Unloading. Shipping containers contain both a liquid and a gas phase of chlorine. The next recommendation assumes the facility has a system to allow for the purging and evacuation of the pipeline/hoses used for loading and unloading.

If the unloading task being done has been periodically undertaken in the past and industrial hygiene sampling results demonstrated that the techniques being utilized may result in chlorine concentrations no more than the capability of the respirator, follow the recommendations outlined here.

Recommendations

- Chlorine gas — Use a full-face respirator approved for protection against chlorine.
- Chlorine liquid — Use a full-face respirator approved for protection against chlorine. Use gloves for thermal (cold) protection.

If the previous criteria have not been met:

- Chlorine gas — Use SCBA or a full-face air supply respirator with an auxiliary self-contained air supply (escape air provision).
- Chlorine liquid — Use SCBA or a full-face air supply respirator with an auxiliary self-contained air supply (escape air provision). Use gloves for thermal (cold) protection.

Emergency Response

The previous recommendations assume that no other hazardous materials requiring additional PPE would be encountered by the responders.

Through their collective experience, the members of the Chlorine Institute have determined that Level B (not enhanced) protection (chemical-resistant clothing) provides appropriate protection to emergency responders for gaseous chlorine releases. Unless it is designed to be self-sealing, the chemical-resistant clothing should be taped using chlorine compatible tape at the openings for the hands and feet. The PPE selected should meet specific criteria that the Institute believes are appropriate for emergency responders to gaseous chlorine as shown in Table 5-3.

The collective experience of the members of the Chlorine Institute reveals that a higher level of protection provides no additional measure of protection to emergency responders of gaseous chlorine releases when not entering a confined space, and, because of its bulkiness, increases the time required to stop the release.

Prior to an individual re-entering a gaseous chlorine release area, such as after the refilling, or replacing any self-contained breathing equipment tank, the individual should be interviewed by a knowledgeable person to verify that skin irritation has not occurred.

Through their collective experience, the members of the Institute have determined that Enhanced Level B, as defined in the following section, provides appropriate protection to emergency responders for liquid chlorine releases.

Recommendations

- Chlorine gas — Follow Level B protection procedures.
- Chlorine liquid — Follow Enhanced Level B protection procedures (as described later in this chapter).

Summary of Recommendations

Tables 5-2 and 5-3 summarize the recommendations contained in this section. The American Water Works Association (AWWA) recognizes that a purchaser of Enhanced Level B PPE may opt to specify Level A PPE in order to reduce the different types of PPE held in inventory or to simplify the PPE selection process.

Enhanced Level B should be used in the following circumstance:

- The hazardous substance has been identified and requires a high level of respiratory protection as in Levels A and B. Skin protection greater than that required by Level B but less than Level A is appropriate. Enhanced Level B protection is appropriate for exposure to several of the chemicals discussed and explained in the following sections.

The Chlorine Institute defines its Enhanced Level B protection as follows:

- The use of a positive-pressure, full-face SCBA or air supply respirator with an auxiliary self-contained air supply (escape air provision).
- The use of protective clothing including
  - — a chemical protective suit,
  - — footwear or footwear cover,
  - — a hood (for protection of head and neck), and
  - — undergarments to provide thermal protection for exposure to liquid chlorine and anhydrous hydrogen chloride.

Enhanced Level B protection provides fully encapsulated protective equipment, but is not gas tight. Level B protection should be used when the

- type and atmospheric concentration of substances have been identified and require a high level of respiratory protection, but less skin protection;
- atmosphere contains less than 19.5 percent oxygen; or
- presence of incompletely identified vapors or gases is indicated, but vapors and gases are not suspected of containing high levels of chemicals harmful to skin or capable of being absorbed through the skin.

NOTE: This involves atmospheres with IDLH concentrations of specific substances that present severe inhalation hazards and that do not represent a severe skin hazard, or that do not meet the criteria for use of air-purifying respirators.

Level C protection should be used when

- the atmospheric contaminants, liquid splashes, or other direct contact will not adversely affect or be absorbed through any exposed skin;
- the types of air contaminants have been identified, concentrations measured, and an air-purifying respirator is available that can remove the contaminants; and
- all criteria for the use of air-purifying respirators are met.

Level D protection should be used when

- the atmosphere contains no known hazards and
- work functions preclude splashes, immersion, or the potential for unexpected inhalation of or contact with hazardous levels of any chemicals.

Table 5-2 Summary of PPE recommendations for tasks involving potential exposure to gaseous or liquid chlorine

	Task Previously Sampled and Within Respiratory Limitations	Task Not Previously Sampled or Sampled and Above Respirator Limitations	
		Gas	Liquid
Initial line break	FFR G	SCBA	Enhanced Level B
Material sampling	FFR G	SCBA	SCBA G
Loading/unloading	FFR G	SCBA	SCBA G
Emergency response	When liquid is not involved = Level B When liquid is involved = Enhanced Level B		

FF = Full-face air-purifying respirator approved for protection against chlorine

G = Gloves for thermal (cold) protection (recommendation is for liquid only)

SCBA = Self-contained breathing apparatus or full-face air supply respirator with an auxiliary self-contained air supply (escape air provision)

Table 5-3 Recommended criteria to evaluate selected PPE components for tasks involving liquid or gaseous chlorine

PPE Component	Enhanced Level B Chlorine Liquid			Level A Chlorine Liquid	Level B Chlorine Gas
	Multi-piece	Hood for Multi-piece	One-piece		
Base material for suit and booties	I and V or I and II	I and V or I and II	I and V or I and II	I and V or I and II	V or II
Visor	N/A	I and V or I and II	N/A	I and V or I and II	N/A
Gloves system	I and V or I and II	N/A	I and V or I and II	I and V or I and II	V or II
Boots	*	N/A	*	N/A	*
Seams/tape	V or II	N/A	V or II	V or II	V or II
Ensemble system	III	N/A	III	IV	N/A

I = ASTM D2136-94 modified to –30°F (low temperature flex text)

II = ASTM F739-99a (chemical resistance—permeation, no breakthrough in 60 min)

III = ASTM F1359-99a (shower test)

IV = ASTM F1052-97 (pressure test)

V = ASTM F903-99a (chemical resistance—penetration; no penetration in 60 min)

N/A = Not applicable

*Boots should be resistant to chlorine and consistent with the facility's foot protection policy. Hood must be used with multi-piece enhanced level.

Emergency Preparedness and Security

Chlorination (and to some extent ammoniation) chemicals and facilities are major components of a total water supply system. A natural or man-made event that affects these components will have an impact on the critical mission of a water supplier. It is essential that a water supplier have an effective emergency preparedness plan that includes these components, is kept current, and incorporates periodic testing and evaluation. Security considerations must be integrated into a system vulnerability assessment to generate an emergency preparedness plan. The plan is a working document that is used before, during, and after a disaster.

The plan is a logical outgrowth of, and developed from, the threat assessment or hazard summary (an example of which is shown in Table 5-4), vulnerability analysis, and implemented mitigation actions.

Regulatory agencies (depending on the jurisdiction) may require specific emergency response plans for hazardous chemicals (chlorine and anhydrous ammonia). Check with your regulatory agency to make sure that all regulatory requirements are satisfied.

A vulnerability assessment for chlorine and ammonia facilities should include, at a minimum, the following components:

- The identification of major system components, including
  - chemical storage,
  - chemical feed systems, and
  - transportation and chemical delivery.
- The identification of threats and potential effects, including
  - natural and
  - man-made.
- The establishment of performance goals and acceptable levels of service.
- The identification of critical system components, including
  - trained personnel,
  - operational considerations,
  - physical facilities,
  - chemical processes,
  - supervisory control and communications, and
  - transportation and delivery.

Once a utility has characterized its critical assets and identified and assessed the threats to those assets, the next step is to define safeguards or mitigation measures that can protect the assets. Such measures include both physical facility improvements and changes in policies or procedures such as personnel training, improved policies and procedures, access control (key control measures, limiting or screening visitors), operational considerations (alternate chemical supplier or standby disinfection strategy), intrusion detection devices, and the use of barriers (fences, traffic restrictions, locks, etc.). Refer to the *Interim Voluntary Security Guidance for Water Utilities* for more information (AWWA 2004).

The emergency preparedness plan should include a description of what must be done if, in spite of the precautions, an event occurs. Every emergency plan must also describe the response in an emergency, a logical recovery plan, and crisis communication procedures. Chlorine and ammonia feed facilities are critical components in a water supply system. An effective emergency preparedness plan can help secure these assets and reduce their vulnerability to identified threats.

Handling Chlorine Emergencies

The presence and use of chlorine can be a potential hazard to both facility employees and the surrounding community. In recognition of this potential, federal law and many state laws require that written emergency plans be developed to prevent, mitigate, and guide response to a chlorine release. There are at least two planning efforts required for each water or wastewater treatment facility: one that addresses protecting the community from a chlorine release and one for protecting employees.

Table 5-4 Hazard summary for a hypothetical water system

Type of Hazard	Estimated Probability	Estimated Magnitude	Comments
Earthquake	1 in 60 years	7.0 (Richter scale)	
Fault rupture	Medium	2 ft	Meridian fault
Ground shaking	High		
Liquefaction	Medium-low	Vertical and horizontal accelerations	Fill areas
Densification	Medium		Fill areas
Landslide	Medium-high		In slopes of 30 percent
Tsunami and seiche	None		
Hurricane	None		
Wind			
Storm surge			
Flooding			
Tornadoes	Low		
Floods	Low-medium	100-year flood to elevation = 1,020 ft	At treatment plant
Forest or brush fires	High		Dry creek watershed
Volcanic eruptions	1 in 300 years	150 miles away	Mount Nueces
Other severe weather			
Snow or ice	None		
Extreme heat	High	100-year drought	Reservoirs depleted
Wind	Medium	60–80 mph	Usually in winter
Lightning	Low		
Other			
Waterborne diseases	Low		Cryptosporidiosis
Hazardous-material release			
Chlorine	Medium-high	1-ton containers	Earthquake damage
Other spill	Medium	Tanker car	Dry creek reservoir
Structure fires	Low		
Construction accidents	Medium	Line damage	In older area of system
Transportation accidents			
Road	Low		
Rail	Medium		Rail yard near warehouse
Water	Low		
Air	Low		
Nuclear power plant accidents	Low	Contamination	Lake West Reservoir
Nuclear bomb explosions	Low		
Vandalism, terrorism	Medium		Storage tanks
Riots	Low		
Strikes	Low		

Before an emergency plan is written, a risk assessment for the facility should be done. Risk assessment is the process of collecting and analyzing information in order to determine what chemical hazards and process risks are present at a facility that could impact employees or the public. Sites with more than 1,500 lb of chlorine in a single process are required to do a risk assessment under the process safety management (PSM) regulations stipulated by OSHA in Section 1910.119 of 29 CFR. The US Environmental Protection Agency (USEPA) requires a risk management plan (RMP) for sites where chlorine exceeds 2,500 lb in a single process, as given in 40 CFR 68. The AWWA Research Foundation developed a generic RMP involving chlorine that could provide information useful in preparing plans suited to specific facilities (Puglionese et al. 1998). RMP planning must include consideration for monitoring, detection, and alarm equipment. Selection of the appropriate emergency personnel, assignment of responsibilities, quantity release estimate, mutual assistance (supplier, hazardous materials [Hazmat] teams, fire departments, etc.), necessary notification requirements (facility and off site), decision making, first-aid needs, and containment should be covered in a set of procedures included in the written plan.

Additional planning considerations should include the technical expertise, scientific instrumentation, heavy equipment, and transportation vehicles that may be needed during an emergency. An inventory of locally available items should be accessible to responders. Likewise, the locations of emergency kits for cylinders, ton containers, cargo tanks, and rail cars should be known. The availability of emergency breathing apparatus, showers, and eye-wash stations and their locations should also be known.

Establishing Procedures

Emergency response procedures are concerned with the efforts of employees from outside the immediate release area or by other designated responders in dealing with an occurrence that results, or is likely to result, in an uncontrollable release of a hazardous substance. Responses to incidental releases of hazardous substances that can be absorbed, neutralized, or otherwise controlled at the time of release by the employees in the immediate release area or by maintenance personnel are not considered to be emergency responses (29 CFR 1910.120).

Each treatment plant should develop its own emergency action checklist, which should be readily available for facility personnel to aid in response. Table 5-5 lists typical key actions that are to be taken in the event of a chlorine emergency. It should be considered as a guide to aid the treatment plant operator. More detailed assistance is available from OSHA, the USEPA, and state regulatory agencies.

Assistance and information during the planning process is available from your chlorine suppliers and the Chlorine Institute (2000). In an emergency situation, responders should be called in the order dictated in your emergency response plan (ERP). These may include the fire service, your chlorine supplier, or another local emergency response team. If you cannot obtain assistance during an emergency, you can contact the Chemical Transportation Emergency Center (CHEMTREC) by calling the toll-free number on your shipping papers. CHEMTREC is designed to assist in transportation emergencies and should be used only as a last resort for assistance.

Training

Training programs and materials are available from a variety of sources, including chlorine suppliers, state and local government agencies, and organizations such as the Chlorine Institute, AWWA, Water Environment Federation (WEF), and the National Fire Protection Association (NFPA). The best starting point for identifying training resources is the local emergency planning committee (LEPC) for your area (contact your state emergency response commission for LEPC information) as well as your supplier.

Table 5-5 Typical emergency action checklist

Action	Performed By	Date/Time
Alert key plant personnel	Discoverer of the problem	
Activate emergency response team if on site	Plant operator	
Determine seriousness of situation	Plant operator	
Ensure that employees are located in a safe area, are properly equipped and protected, and that all are accounted for	Plant operator	
Alert appropriate off-site authorities	Plant operator	
Ensure that employees who may have been exposed receive medical surveillance and treatment if necessary	Plant operator	
Attempt to stop or control release	Response team	
Obtain outside assistance (via the supplier or CHLOREP via CHEMTREC at: 1-800-424-9300 [US]; 1-703-527-3887 [Alaska, Hawaii, D.C.]; 1-613-996-6666 [Canada; collect call])	Team	
Bring the incident under control	On-site coordinator	
Determine when normal operations can resume	On-site coordinator	
Provide close-out report of incident	On-site coordinator	

NOTE: Fill in last column with your information.

Facility personnel. Treatment plant employees should be trained in the emergency response plan, in safety procedures for the handling and use of chlorine, and in the use of self-contained breathing apparatus and other applicable equipment. The training requirements depend on the specific employee's roles and responsibilities. Each plant should have a training program customized to its specific needs depending on the type of facility, type and number of chlorine containers, and number of employees. The site should keep written documentation of all training.

Emergency responders. Emergency responder training requirements are based on the response level and type of job responsibility assigned to each responder. The following list includes the various response levels of those who may be present at the site of an emergency and brief descriptions of their responsibilities:

- *First Responder, Awareness Level:* Individuals who, in the course of their normal duties, may be first on the scene of an emergency involving a hazardous substance. They are expected to notify the proper authorities as indicated in the plant's emergency response plan and take no further action.
- *First Responder, Operations Level:* Individuals involved in the initial response to a release or potential release of hazardous substances for the purpose of protecting nearby persons, the environment, or property from the effects of the release. They are trained to respond defensively without actually trying to stop the release. Their function is to contain the release from a safe distance, keep it from spreading, and prevent exposures.
- *Hazardous Materials Technician:* Individuals who respond to a release or potential release for the purpose of stopping the release. They assume a more aggressive role than first responders at the operations level do in that they will approach the point of release and attempt to plug, patch, or otherwise stop the release. Appropriate Chlorine Institute emergency kits, shown in Figure 5-2, must be available for applicable gas storage containers (A for cylinders, B for ton containers, C for tank cars).
- *Hazardous Materials Specialist:* Individuals who respond with and provide support to hazardous materials technicians. Their duties parallel those of the

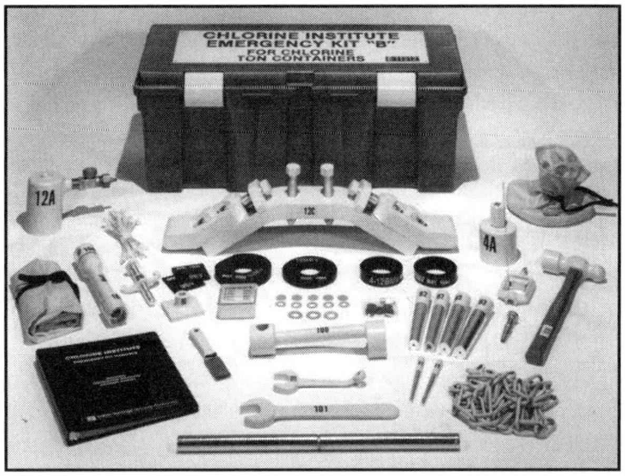

Source: The Chlorine Institute.

Figure 5-2 Chlorine Institute emergency kit B

Table 5-6 Summary of Hazmat responders training requirements

Response Level	Minimum Training Requirement
Awareness level	Understanding hazardous materials, including their risks, and how to secure the site and notify others in case of an emergency
First responder, operations level	8 hr of training, including awareness-level topics
Hazardous materials technician	24 hr of training equal to technician-level topics
Hazardous materials specialist	24 hr of training equal to operational level plus competency in commanding incidents and implementing emergency response plans
Skilled support personnel	Initial pre-entry briefing
Specialist employees	Annual training and competency in area of specialization

technicians but require a more specific knowledge of the various substances they may be called on to contain. The specialist also may act as the site liaison with federal, state, local, and other government authorities regarding site activities.

- *On-Scene Incident Commander:* The individual who is responsible for directing and coordinating all aspects of a hazardous incident.
- *Skilled Support Personnel:* Individuals skilled in operating certain equipment (e.g., backhoe or crane) and are needed temporarily to perform immediate emergency support work.
- *Specialist Employees:* Individuals who, in the course of their regular job duties, work with and train in the hazards of specific hazardous substances, and who may be called on to provide technical advice or assistance.

It is beyond the scope of this manual to provide the details of a training program for emergency responders; however, a summary of the training requirements is provided in Table 5-6. Actual training requirements for each level of responder are given in 29 CFR 1910.120. These regulations change, so 29 CFR should be reviewed periodically.

Audits and Exercises

An effective way to determine the adequacy of a treatment plant emergency plan is to have periodic audits and exercises. Audits should be performed with various facility response personnel to test their knowledge of duties and equipment, along with periodic auditing on actual use of the equipment. Exercises should be conducted to test the participants' reactions and effectiveness in implementing the emergency plan as well as to test the actual mechanics of the plan.

There are three types of exercises: the full-scale exercise, the in-plant exercise, and the tabletop exercise. Consideration should be given to conducting full-scale exercises utilizing responders from the community at least once a year. Periodic in-plant exercises should use different simulated events and involve as many of the various personnel as possible. These exercises should be conducted similarly to full-scale exercises but would not involve outside emergency personnel.

Tabletop exercises should be conducted periodically to check the ability of the emergency response crews to analyze an event, communicate effectively to outside emergency response personnel, and respond to unfolding events. This type of exercise is usually conducted with just the supervisors of key emergency response personnel, both in-plant and from outside agencies.

Following any of the exercises, a critique should be made to assess the effectiveness of the plan and to pinpoint any weaknesses in it or in the training and knowledge level of the personnel involved. A written report of the exercise should be available for review and the facility's emergency plan should be modified as needed.

MEDICAL ASPECTS AND FIRST AID—CHLORINE

Chlorine gas is primarily a respiratory irritant. It is so intensely irritating that low concentrations (0.02/0.2 ppm) are readily detectable by most people. At low concentrations chlorine gas has an odor similar to household bleach. As the concentrations increase from the level of detection by smell, so does the symptomatology in the exposed individual. At chlorine concentrations above 5 ppm, the gas is very irritating, and it is unlikely that any individual would remain in such an exposure for more than a very brief time unless trapped or unconscious. The effects of exposure to chlorine may become more severe up to 36 hr after the incident. Therefore, close observation of exposed individuals should be a part of the medical program. For additional information, see Chlorine Institute pamphlets 63 and 90, Section 1.2 (Chlorine Institute 1998a, 1998b), and the Chlorine Institute video, *Health Effects from Short-Term Chlorine Exposure*.

Acute Toxicity

Chlorine gas. In concentrations near the threshold of smell, chlorine gas will cause mild irritation of the eyes and of the mucous membrane of the respiratory tract after several hours of exposure. As concentrations increase, there is an increase in the irritating effect on the eyes, coughing mechanism, and upper and lower respiratory tract with eventual difficulty in breathing. As the duration of exposure or the concentration increases, the affected individual may become apprehensive and restless, with coughing accompanied by throat irritation, sneezing, and excess salivation. At higher levels, there is vomiting associated with labored breathing. In extreme cases, difficulty in breathing may progress to the point of death through suffocation. An exposed individual with a pre-existing medical or cardiovascular condition may have an exaggerated response.

Liquid chlorine. Liquid chlorine in contact with the eyes or skin will cause local irritation and/or burns. All symptoms and signs result directly or indirectly from its direct irritating action. There are no known systemic effects.

Chronic Toxicity

Most studies indicate no significant connection between adverse health effects and chronic exposure to low concentrations of chlorine. A Finnish study demonstrated an increase in chronic coughs and a tendency for hypersecretion of mucus among workers (Grenquist-Norden 1983, in the *Chlorine Manual* 1996.) However, these workers showed no abnormal pulmonary function in tests or chest X-rays.

The Chemical Industry Institute of Toxicology issued its report on a study of the chronic inhalation of chlorine in rats and mice. The animals were exposed to chlorine gas at 0.4, 1.0, or 2.5 ppm for up to 6 hr a day and for 3 to 5 days a week for up to 2 years. There was no evidence of cancer. Exposure to chlorine at all levels produces nasal lesions. Because rodents are obligatory nasal breathers, how these results should be interpreted for humans is not clear and is under investigation (Chemical Industry Institute of Toxicology, 1993, in the *Chlorine Manual* 1996).

Preventive Measures

The Chemical Industry Institute of Toxicology recommends that baseline and periodic medical examinations be considered for employees that may be subjected to chlorine exposure. The examinations should consist of a complete medical history and physical examination, including chest X-ray and baseline respiratory function studies. Specific reference to respiratory allergies, congenital or acquired pulmonary disease such as asthma or emphysema, cardiac disease, and chronic eye conditions should be ascertained.

Water treatment plants should adopt a medical surveillance program suitable to their needs. It should be determined that the employee is physically capable of wearing respiratory protection equipment.

First Aid

First aid is the immediate temporary treatment given to an exposed individual before the services or recommendations of a physician are obtained. Prompt action is essential. Firmness and assurance will help to alleviate anxiety. Medical assistance must be obtained as soon as possible. Never give anything by mouth to an unconscious or convulsing individual. **Caution:** The individual providing first aid must use proper personal protective equipment to avoid becoming a casualty.

Inhalation

Respiratory assistance.

If an exposed individual has stopped breathing:

- Move the individual to fresh air.
- Remove and dispose of any contaminated clothing.
- Call for emergency medical assistance.
- Begin cardiopulmonary resuscitation (CPR) immediately.
- Administer humidified oxygen by trained personnel as soon as possible.

If an exposed individual is still breathing:

- Move the individual to fresh air.
- Remove and dispose of any contaminated clothing.
- Place the individual in a comfortable position in a chair; in severe cases, have the individual lie down with his or her head and trunk elevated to a 45–60° angled position.
- Encourage slow, regular breathing.

- Administer humidified oxygen by trained personnel as soon as possible.
- Keep the individual warm and at rest.
- Render any other necessary first aid.

Alleviate anxiety by explaining the various procedures being used, and obtain cooperation, especially for breathing exercises.

Oxygen administration. Oxygen should be administered by first-aid attendants trained in the use of the specific oxygen equipment on hand, either on site or at a nearby facility. Such equipment should be periodically tested. More sophisticated inhalation equipment is available in most emergency facilities. Humidified oxygen should be used whenever possible.

Skin contact

Liquid chlorine

If liquid chlorine has contaminated the skin or clothing, an emergency shower should be used immediately, and contaminated clothing should be removed while under the shower. Flush contaminated skin with copious amounts of running water for 15 min or longer. Thermal burns, due to the cold temperature of liquid chlorine, may be more damaging than any chemical reaction of chlorine and the skin.

Chlorine gas

Exposure to gaseous chlorine may irritate the skin. Do not attempt neutralization or apply any salves or ointments to damaged skin. Refer the individual to a physician if irritation persists after irrigation or if the skin is broken or blistered.

Eye contact. If the eyes have been exposed to any concentration of chlorine in excess of the PEL, they should be flushed immediately with copious quantities of tepid running water or a direct stream of water for at least 15 min. **Caution:** When dealing with exposure to the eyes, never attempt to neutralize with chemicals.

The eyelids should be held apart while being flushed to ensure contact of the water with all accessible tissues of the eyes and lids. Medical assistance must be obtained as soon as possible. If such assistance is not immediately available, eye irrigation should be continued for a second 15-min period. Nothing but water should be applied unless ordered by a qualified healthcare provider.

Excitement phenomenon. Exposure to chlorine can lead to excited behavior as a result of central stimulation and emotional disturbances. Reassurance is best accomplished without the use of sedatives. The use of sedatives should only be considered by qualified medical personnel following medical assessment and only under close supervision of respiratory function to monitor progress.

Delayed effects. The inhalation of any chlorine may lead to delayed reactions such as pulmonary edema. Because physical exercise appears to have some relation to the incidence of delayed reaction, it is recommended that any patient who has had severe inhalation exposure should be kept at rest for a period of observation. The length of observation will depend on the clinical assessment of the exposed individual. Observation may be required for up to several days after the exposure. Excitement, apprehension, or emotional distress may persist after a period of observation following severe exposure.

SODIUM HYPOCHLORITE SAFETY CONSIDERATIONS

All persons who distribute, store, or use sodium hypochlorite (NaOCl) solution should have in their possession and be thoroughly familiar with the material safety data sheet (MSDS). The product's MSDS contains information regarding its chemical characteristics, physical hazards, health hazards, first aid, transportation information, fire fighting information, and environmental information. Contact a manufacturer of sodium hypochlorite solutions for assistance in obtaining this information.

Sodium hypochlorite will also react with acids, ammonium hydroxide (ammonia water), or cleaners containing ammonia compounds to produce hazardous gases. Sodium hypochlorite solutions may also react violently with some organic compounds. Care must be taken to prevent mixing of these noncompatible compounds.

Sodium hypochlorite decomposes when exposed to heat. If sodium hypochlorite solutions are contaminated, a vigorous reaction along with a pressure build-up is possible if the product decomposes in a container. Vented closures may be used when storing higher-strength sodium hypochlorite solutions to avoid excessive pressure build-up. Care must be taken when opening these containers to make certain excess pressure is not present. Although spraying of the product is unlikely, eyes and skin must be protected if the container is under pressure.

Sodium hypochlorite may cause damage to skin and eyes. Therefore, proper safety equipment must be worn when handling sodium hypochlorite solutions. Such equipment could include the use of respiratory equipment when necessary (if fumes are present), goggles, and impervious gloves, boots, and apron. (For more information see the "Personal Protective Equipment" sections in this chapter.) For household bleach (6 percent or lower) refer to label directions for proper handling instructions.

Precautions for Handling Industrial Strength Sodium Hypochlorite Solutions

Precautions for handling industrial strength sodium hypochlorite solutions include the following:

- Keep the container closed when not in use.
- Ensure adequate ventilation or use an approved appropriate respiratory protection mask.
- Avoid breathing fumes.
- Avoid contact with eyes, skin, and clothing.
- Wash thoroughly after handling.

Wear appropriate personal protective equipment, which may include goggles and face shield; impervious gloves, boots, and apron; or suit when handling this product, and follow these additional precautions:

- Do not allow contact with organic materials (e.g., rags, wood fibers, paper debris, etc.) or reducing chemicals except under controlled conditions. Do not discard indiscriminately. A spontaneous combustion (fire) could result.
- Do not mix with acids, ammonia, or reducing agents. To do so may release hazardous gases.

Precaution of Handling Household Strength Bleach

When handling household strength bleach, it is important to read and follow the instructions on the label and use the safety equipment recommended.

Physiological Effects of Sodium Hypochlorite (3–20 wt%)

Sodium hypochlorite solutions are classified as a corrosive by the US Department of Transportation (USDOT), and they are a mild-to-severe irritant to the eyes, skin, mucous membranes, and the respiratory system. Exposure may occur by either direct contact with sodium hypochlorite solutions or entrained mists and aerosols.

The primary concerns with exposure to sodium hypochlorite solutions are with the eyes, followed by the mucous membranes, the respiratory system, and the skin. The impact of exposure to sodium hypochlorite is dependent on the concentration of the solution, the amount of excess sodium hydroxide contained in the solution, and the contact time. Excess sodium hydroxide in the solution may cause severe irritation and/or burns and possible blindness. Sodium hypochlorite solutions, mists, or aerosols may also cause skin irritation. The severity may be reduced by prompt flushing of the affected areas with copious amounts of water and obtaining immediate medical attention.

Ingestion of sodium hypochlorite solution may cause severe burns to the mucous membranes of the mouth, throat, esophagus, and stomach. Breathing mist or spray may cause damage to the upper respiratory tract and lungs, which could lead to chemical pneumonia, depending on the severity of the exposure.

Neither a PEL nor TLV has been established. The American Industrial Hygiene Association (AIHA) recommends a workplace environmental exposure level (WEEL) guide for sodium hypochlorite solutions at 2 mg/m^3, 15-min time-weighted average exposure. The predominant chlorine-like odor associated with sodium hypochlorite is hypochlorous acid for which there are no established exposure limits. The exposure limits for chlorine may be considered as applicable in many circumstances.

Personal Protective Equipment Selection for Sodium Hypochlorite (3–20 wt%)

The following recommendations assume that the facility has not performed a detailed JSA of the specific task being performed. If such a detailed JSA of the specific task has been performed and documented, and it concludes that a different level of PPE will protect the employee(s) performing the work, such different levels of PPE are compatible with the purposes and intent of these recommendations.

This section covers the recommended PPE for performing the specified tasks involving sodium hypochlorite at concentrations between 3 and 20 percent at temperatures below 100°F (38°C). Typically, sodium hypochlorite must be maintained below 100°F (38°C) to minimize decomposition. Materials for PPE should be chemically resistant against 3–20 percent sodium hypochlorite at 100°F (38°C) or the applicable temperature and meet the criteria in Tables 5-7 and 5-8. Recommendations for specific activities are provided here.

NOTE 1: Respiratory equipment may be needed in situations during which there is a risk of sodium hypochlorite mixing with acidic or other incompatible materials resulting in the release of chlorine gas.

NOTE 2: When chemical protective equipment is worn to protect the feet and body, and the garment has pant legs but does not have integral foot protection, the legs of the protective garment must be placed on the outside of the protective footwear.

Initial line break: Chemical protection for the head, face, eyes, hands, body, and feet.

Material sampling: Chemical protection for the face, eyes, and the hands.

Unloading: Unloading a container typically involves either pressurizing the container or installing connections at a valve located on the bottom of the container or inserting a pump into the container.

While inspecting an open dome when product is flowing:

- Chemical protection for the head, face, eyes, hands, body, and feet.

While inspecting an open dome when no product is flowing:

- Chemical protection for the eyes.

Emergency Response

The recommendations outlined here assume that no other hazardous materials requiring additional PPE would be encountered by the responders. Responders should also be aware that sodium hypochlorite will react with acidic and other incompatible materials (e.g., ammonia, organics), resulting in the release of chlorine or other hazardous chemicals or a fire. In such situations, obtain chemical protection for the head, neck, face, eyes, hands, body, and feet. Respiratory protection recommendations are as follows.

In severe cases with spraying sodium hypochlorite in a major leak:

- SCBA or full-face air supply respirator with an auxiliary self-contained air supply (escape air provision).

In less severe cases such as a leaking valve or pipeline with no appreciable spraying and/or splashing product:

- No respiratory protection is needed.

This recommendation is based on the assumption that unless otherwise determined by the incident commander, the responders to a sodium hypochlorite release will not be exposed to concentrations of fumes or material in excess of 2 mg/m^3 in any 15-min period.

Summary of Recommendations

Tables 5-7 and 5-8 summarize the recommendations contained in this section.

Table 5-7 Summary of PPE recommendations for tasks involving potential exposure to 3–20 percent sodium hypochlorite below 100°F

	Chemical Protective Hat or Hood	Face Shield and Chemical Splash Goggles	Chemical Splash Goggles (Without Face Shield)	Chemical Protective Suit	Chemical Protective Gloves	Chemical Protective Boots or Over-shoes	Respiratory Protection
Initial line break	R	R		R	R	R	N/A
Material sampling	N/A	R		N/A	R	N/A	N/A
Loading (remotely activated)	R	R		N/A	R	N/A	N/A
Loading (not remotely activated)	R	R		R	R	R	N/A
Unloading	R	R		R	R	R	N/A
Loading/unloading (only when inspecting dome with no product flowing)	N/A	N/A	R	N/A	N/A	N/A	N/A
Emergency response*							
Severe cases (spraying)	R	R		R	R	R	SCBA
All others	R	R		R	R	R	N/A

R = Recommended PPE for this task

N/A = This PPE is not believed necessary for this task

SCBA = SCBA or full-face air supply respirator with an auxiliary self-contained air supply (escape air provision)

*Chemical protection of the neck (e.g., hood) is also recommended for emergency response

Table 5-8 Recommended criteria to evaluate PPE components for tasks involving 3–20 percent sodium hypochlorite

PPE Component	Recommended Test
Base material, seam, visor, gloves, boots, overshoes	V or II performed at the applicable temperature
Ensemble system	III

II = ASTM F739-99a (chemical resistance: permeation, no breakthrough in 60 min)

III = ASTM F1359-99a (shower test)

V = ASTM F903-99a (chemical resistance: penetration; no penetration in 60 min)

NOTE: The appropriate shoe csonsistent with the facility's foot protection policy should be worn.

First Aid

Eyes. Immediately flush with large amounts of water for at least 15 min, occasionally lifting the upper and lower eyelids. Seek medical attention at once.

Skin. Immediately flush with water for at least 15 min. Seek medical attention. If clothing, shoes, and/or jewelry comes in contact with the product, they should be removed immediately and laundered before re-use.

Inhalation. If an individual experiences nausea, headache, or dizziness, the individual should stop work immediately and move to fresh air until these symptoms disappear. If breathing is difficult, and if properly trained, administer moist oxygen, and keep the individual warm and at rest. Seek medical attention. In the event that an individual inhales enough vapor to lose consciousness, the individual should be moved to fresh air at once and seek medical attention immediately. If breathing has stopped, artificial respiration should be given immediately. In all cases, ensure adequate ventilation and provide respiratory protection before the individual returns to work.

AMMONIA GAS SAFETY CONSIDERATIONS

Ammonia is not a cumulative poison. The effects of exposure to ammonia vapor depend on exposure concentration and time of exposure. The effects may vary from mild irritation to breathing difficulties to edema and severe damage of the mucous membranes with possible fatal results. Individuals can have different responses to exposure. Those with chronic respiratory diseases should avoid exposure to ammonia. Table 5-9 lists the typical physiological effects.

Ammonia is highly alkaline when combined with water. Caustic burns may result from contact with skin or mucous membranes. Ammonia may absorb water from an area thus causing dehydration of tissues.

Liquid ammonia boils at –28°F (–33°C) under normal atmospheric conditions. Liquid ammonia, thus, may cause frostbite if contacted with skin. Some of the exposure guidelines for ammonia are listed as follows:

Ammonia exposure guidelines (mg/L)

OSHA Permissible Exposure Limit	50
NIOSH Immediately Dangerous to Life or Health Concentration	300
NIOSH Recommended Exposure Limit	25
ACGIH Threshold Limit Value	25

Personal Protective Equipment

Safety regulations and local requirements may dictate the use of certain protective equipment for workers handling ammonia. The suggestions listed here do not replace

Table 5-9 Physiological effects of ammonia

Ammonia (ppm)	Effect
5	Least perceptible odor
20–50	Readily detectible odor
50–100	No discomfort or impairment for prolonged exposure
150–200	General discomfort
400–700	Severe irritation of eyes, ears, and nose
1,700	Coughing, bronchial spasms
2,000–3,000	Dangerous, may be fatal
5,000–10,000	Serious edema, rapidly fatal
10,000+	Immediately fatal

Source: Compressed Gas Association 1992.

these requirements but are offered as guidelines for consideration by operators of ammonia storage and feed facilities.

Chemical splash-proof goggles and rubber (or plastic) gauntlet gloves rated for use with ammonia should be worn by workers under routine conditions. A full-face shield may be worn over the goggles but not instead of the goggles.

Additional protective and safety equipment must be provided for use in an emergency (e.g., leak or spill). This equipment may include one or more of the following:

- Safety shower and/or eye wash
- Respiratory devices (NIOSH approved)

  Full-face gas mask with ammonia or universal NIOSH-approved canister should be used. Canisters are limited to brief periods of use not exceeding 15 min and concentrations of ammonia in air by volume not exceeding 20,000 ppm for escape or 300 ppm for entry purposes. An ammonia-approved supplied air respirator with full-face piece may also be used where allowed.

Use SCBA with full-face piece for protection when entering an area where the ammonia concentration is unknown or exceeds the IDLH level of 300 ppm or in oxygen-deficient atmospheres. User must be OSHA trained in the proper use of this equipment.

- Protective clothing
- Rescue harness
- High-capacity water system
- Stretcher and blankets

First-aid Procedures

Ammonia is very water soluble. First aid for contact with ammonia is to flush the affected area with large quantities of clean water. It is generally recommended that the skin or eyes receive continuous application of water for at least 15 min to reduce injury. An individual should receive treatment by a physician if he or she has been burned or overcome by ammonia. The first-aid procedures presented here are believed to be common practice in the industry. Adopting these procedures in a specific case should be subject to prior endorsement by a competent medical advisor.

Inhalation. Conscious individuals exposed to ammonia vapor should proceed at once to a location free of ammonia. If exposure is minimal, no additional treatment may be necessary.

If an individual is overcome by ammonia, he or she must be removed to a location free from ammonia. Seek the services of a physician immediately. If resuscitation is necessary, the procedure must be applied with all haste.

Eyes. Flood with copious quantities of clean water. Remove contact lenses if they are worn. Seek the services of a physician for further treatment.

Skin and mucous membranes. Flood with copious quantities of clean water. Seek the services of a physician for further treatment.

Internal. Swallowing liquid ammonia is very unlikely. If conscious, have the individual drink large quantities of water immediately. Transport the individual to a physician and seek his or her advice promptly.

Ammonia Gas Emergencies

An emergency preparedness and response plan should include procedures and actions to address an ammonia leak or spill. Regulations may require the inclusion of certain specific elements in an emergency plan. Some items to consider when preparing an emergency plan for an ammonia release include the following:

- Ensure personnel are trained to control a leak or contain a spill.
- Wear PPE as required by the situation.
- Stay upwind of the leak if possible.
- Be aware that irritating vapor generally will force personnel to evacuate before being overcome.
- Ensure that trapped personnel stay close to the floor because ammonia vapor is lighter than air.
- Warn individuals in the path of a liquid ammonia spill (extensive spill) and notify authorities to initiate an evacuation plan.
- Make sure good ventilation is available, as it can be effective in dissipating ammonia, which is lighter than air.
- Do not attempt to neutralize an ammonia spill with an acid.
- Dilute a spill with at least 100 times the volume of water. Take care to avoid an additional problem that may be caused by the runoff.
- Make sure there is an up-to-date telephone list of emergency numbers.

Leak detection. Most ammonia leaks are readily detected by ammonia's pungent odor. There are various types of devices to detect and measure ammonia vapor in air. Both continuous and instantaneous detection devices are available. Continuous toxic gas detectors generally use electro-chemical cell sensors that detect a wide range of gases, including ammonia. Instantaneous detection devices may use colorimetric comparison or electro-chemical sensors.

Leak control. Most leaks can be controlled or stopped by closing the supply valve to the system or tightening valve stem packing gland nuts. Storage tanks may develop leaks at threaded or flanged fittings. Careful tightening of these threads or bolts may stop the leak. The supplier should be contacted for assistance in dealing with damaged or leaking shipping containers. If the producer or supplier cannot be reached, contact CHEMTREC for advice.

Fire exposure. Remove the ammonia container from the fire zone if possible. If the container cannot be moved, it should be kept cool with water until after the fire is extinguished. Fire-fighting personnel should be equipped with protective clothing and respiratory equipment.

AQUA AMMONIA (AMMONIUM HYDROXIDE) SAFETY CONSIDERATIONS

Ammonium hydroxide is a caustic solution that can cause injury to individuals upon contact. In addition, ammonia gas is readily released from ammonium hydroxide solutions depending on the concentration and temperature of the liquid solution. Therefore, safety precautions must be taken to protect individuals from the caustic solution as well as the ammonia gas. (For information about ammonia gas safety procedures and equipment, see "Ammonia Gas Emergencies" earlier in this chapter.) Suppliers must provide an MSDS for ammonium hydroxide, which should be consulted for specific hazard and safety information. Also, all applicable regulations must be followed. The information provided in this section contains safety information that follows general accepted practices. This information is not meant to substitute for regulatory requirements or physicians' instructions.

Personal Protective Equipment

In general, the same equipment is needed for aqua ammonia as for ammonia gas protection. The ammonium hydroxide solution is caustic and can cause severe burns to skin and damage mucous membranes. Chemical protective clothing (e.g., gloves, coveralls, boots, etc.) that can resist caustic ammonium hydroxide is particularly important. Eye protection (e.g., chemical goggles) is required when handling this material. Contact lenses should not be worn because they may contribute to severe eye injury if the solution contacts the eyes.

First-aid Procedures

The first-aid procedures for ammonium hydroxide are essentially the same as listed for ammonia gas (listed earlier). This is due to the fact that ammonia gas is very soluble in water and thus many of the hazards of the gas are associated with the formation of ammonium hydroxide on the skin or mucous membranes. Ammonium hydroxide solution is caustic and can cause severe burns with deep ulceration and permanent scarring. If swallowed, it can cause burns to the mouth, throat, and esophagus, which may progress to stricture formation. Flush the area contacted with the solution with large amounts of water. In the unlikely event that the solution is ingested, have the victim drink a large quantity of water. In either case, transport the victim to an emergency care facility immediately.

Aqua Ammonia Emergencies

An emergency preparedness and response plan should include procedures and actions to address an aqua ammonia leak or spill. Regulations may require the inclusion of certain specific elements in an emergency plan. Some items to consider when preparing an emergency plan are listed in "Ammonia Gas Emergencies" discussed earlier.

Engineering Controls

The goal is to reduce hazardous exposure. Methods include mechanical ventilation, process or personnel enclosure, control of process conditions, and process modification. Personal protective equipment is also required for persons handling the material. Enclosure of the storage and feed equipment is recommended to isolate the solution to limit exposure potential. For large-scale handling operations, use non-sparking, corrosion-resistant ventilation systems, approved explosion-proof equipment, and intrinsically safe electrical systems. Ammonia leak detection systems with alarms may be required or desired. Walls, floors, shelving, fittings, lighting, and ventilation systems should be made from carbon steel or stainless steel that does not react with ammonia or ammonium hydroxide. Supply sufficient replacement air to make up for air removed by exhaust systems. Emergency showers and eyewash stations should be provided in close proximity.

Leak Control and Containment

Most leaks develop in storage tanks and feed systems. Leaks are easily detected by the pungent odor of ammonia or detection sensors. This material is toxic and corrosive and readily gives off ammonia gas, which is also an explosion hazard in confined spaces. Maintenance and emergency personnel should immediately put on required personal protective equipment including respirator protection before attending to a repair of the leak. Closed handling systems should be considered for large-scale facilities. Consideration should be given to adequate containment and possible neutralization for bulk storage tanks. Follow handling precautions on the MSDS. Suitable emergency equipment for fires, spills, and leaks must be readily available. Storage of ammonium hydroxide must be in accordance with applicable occupational health and safety regulations and local fire and building codes. Notify government environmental agencies if there is a release of ammonium hydroxide into the environment.

The supplier should be contacted for assistance in dealing with damaged or leaking shipping containers. If the producer or supplier cannot be reached, contact CHEMTREC for advice.

Fire-Fighting Precautions

Evacuate the area and fight fire from a safe distance. Approach the fire from upwind to avoid hazardous gases and toxic decomposition products. Move containers from the fire area if it can be done without risk. Explosive decomposition may occur under fire conditions. Fire-exposed containers should be cooled by application of hose streams. Ammonium hydroxide and ammonia gas are corrosive. Special protective clothing may be necessary to provide adequate protection. Refer to the MSDS for detailed information on fire-fighting instructions.

Special Consideration for Facilities With Hypochlorite Storage

Steps should be taken to prevent accidentally unloading ammonium hydroxide into a liquid sodium hypochlorite tank (or the reverse). An explosive mixture of nitrogen trichloride may be produced. Nitrogen trichloride is also a toxic compound and may cause a hazardous situation if released into the air.

CHLORINE AND AMMONIA FACILITY REQUIREMENTS

Recommended Standards for Waterworks (Ten States Standards)

Each state may have its own storage and use requirements as well as those recommended by the various industry associations (e.g., the Chlorine Institute, Compressed Gas Association, and Chemical Manufacturers Association) and the federal government (OSHA and USEPA) regarding the storage and use of these chemicals. It is incumbent on the user to determine the applicable requirements. In addition, common sense should prevail and any errors should be made in the direction of safety.

Arguably one of the most focused sets of standards for water industry use is that promulgated by the Great Lakes–Upper Mississippi River Board of State Public Health and Environmental Managers, commonly called the *Ten States Standards*. These standards are written specifically for the water industry and are used as the guide for chlorine, ammonia, and chemical feed installations. The *Ten States Standards* were developed by representatives from the following 10 states and 1 Canadian province: New York, Pennsylvania, Ohio, Indiana, Michigan, Illinois, Wisconsin, Iowa, Minnesota, Missouri, and Ontario. The standards are updated from time to time to meet new developments. Some of the requirements from the *Ten States Standards* are as follows:

- A minimum of two feeders shall be provided (one for standby).
- When booster pumps are used, a standby pump shall be provided.
- Separate feeders shall be used for each chemical provided.
- Automatic controls, when provided, shall have a manual override.
- Feeders shall have an electrical interlock with the well or service pump.
- Chemical feed rates shall be proportional to flow.
- Scales shall be provided for measuring the quantities of chemicals used (see Figures 4-9 and 4-10).
- Liquid chemical feeders shall have antisiphon protection.
- Water supply to feeders shall have cross-connection protection.
- Chemical feed equipment shall be located in rooms separate from the rest of the facility.
- Space should be provided for 30 days of chemical storage.
- Day tanks should hold no more than a 30-hr supply.
- Day tanks shall be provided with a calibrated gauge.
- Respiratory masks of the self-contained, air supply type shall be provided.
- Ammonia solution for chlorine leaks and hypochlorite solution for ammonia leaks shall be provided.
- Chlorine Institute emergency kits shall be provided when using ton containers.
- When a leak detector is provided, it shall have both an audible alarm and warning light.
- Protective equipment consisting of rubber gloves, an apron or similar clothing, goggles or face mask, deluge shower, and eye-washing device shall be provided.
- Chlorine cylinders should be isolated from operating areas, chained in an upright position, and stored in rooms separate from ammonia.
- Chlorine and ammonia rooms shall be enclosed, separated from other rooms, provided with a shatterproof inspection window, provided with doors that have external exits only, and have doors equipped with panic hardware.
- Chlorine and ammonia rooms shall have ventilating fans with a capacity of one complete air change per minute; air inlets shall include louvers near the floor; and air discharge shall be above grade, not near walkways, doors, or fresh-air intakes.
- Chlorine and ammonia rooms shall be provided with ventilation.
- Chlorine and ammonia rooms shall be heated to 60°F (16°C) and pressurized chlorine gas shall not be carried beyond the chlorinator room.
- Chlorine rooms shall not be below ground level.

Although this list appears lengthy, it is by no means complete. It is recommended that interested parties obtain a copy of the *Ten States Standards* for review. These standards have no legal standing unless the state codes require their use or have adopted them in some form.

REFERENCES

American Conference of Industrial Hygienists (ACGIH). 2001. *Documentation of the Threshold Limit Values and Biological Exposure Indices,* 7th ed. Cincinnati, Ohio: American Conference of Industrial Hygienists.

CFR (Code of Federal Regulations). Available at: http://www.gpoaccess.gov/cfr/index.html.

Chlorine Institute. 1996. *Chlorine Manual,* 6th ed. Arlington, Va.: The Chlorine Institute.

———. 1998a. *First Aid and Medical Management/Surveillance and Occupational Hygiene Monitoring Practice for Chlorine.* Arlington, Va.: The Chlorine Institute.

———. 1998b. *Molecular Chlorine: Health and Environmental Effects.* Arlington, Va.: The Chlorine Institute.

———. 2000. *Emergency Response Plans for Chlorine Facilities.* Arlington, Va.: The Chlorine Institute.

———. 2001. *Personal Protective Equipment for Chlor-Alkali Chemicals.* Arlington, Va.: Chlorine Institute.

CGA (Compressed Gas Association). 1984. *Anhydrous Ammonia, Pamphlet G-2.* Alexandria, Va.: Compressed Gas Association.

Puglionese, P.S., H.S. McGee, J. Tittensor, and J.O. Hammell. 1998. *Compliance Guidance and Model Risk Management Program for Water Treatment Plants.* Denver, Colo.: Awwa Research Foundation.

ADDITIONAL SOURCES OF INFORMATION

ANSI (American National Standards Institute). 1999. *ANSI-K61.1, Safety Requirements for the Storage and Handling of Anhydrous Ammonia.* New York: ANSI.

ASCE (American Society of Mechanical Engineers). 1995. Section VIII, Rules for Construction of Pressure Vessels. In *Boiler and Pressure Vessel Codes.* New York: ASME.

ASCE and AWWA (American Society of Civil Engineers and American Water Works Association). 1997. *Water Treatment Plant Design,* 3rd ed. New York: McGraw-Hill.

AWWA. 2003. *Principles and Practices of Water Supply Operations: Water Treatment,* 3rd ed. A.M. Wachinski, ed. Denver, Colo.: AWWA.

Chlorine Institute, Inc. 1995. 2002. *Sodium Hypochlorite Handling and Storage Guidelines.* Pamphlet 100, 3rd ed. Arlington, Va.: The Chlorine Institute, Inc.

———. 1997a. *The Chlorine Manual.* Pamphlet 1, 6th ed. Arlington, Va.: The Chlorine Institute, Inc.

———. 1997b. 2002. *Chlorine Vaporizing Equipment.* Pamphlet 9, 5th ed. 6th ed. Arlington, Va.: The Chlorine Institute, Inc.

———. 1998a. *Chlorine Scrubbing Systems.* Pamphlet 89, 2nd ed. Arlington, Va.: The Chlorine Institute, Inc.

———. 1998b. *Communicating Your Chlorine Risk Management Plan.* Arlington, Va.: The Chlorine Institute, Inc.

———. 1998c. 2005. *Piping Systems for Dry Chlorine.* Pamphlet 6, 14th ed. 15th ed. Arlington, Va.: The Chlorine Institute, Inc.

———. 1999. *Water and Wastewater Operators Chlorine Handbook.* Pamphlet 155. Arlington, Va.: The Chlorine Institute, Inc.

———. 2000. *Sodium Hypochlorite Manual.* Pamphlet 96, 2nd ed. Arlington, Va.: The Chlorine Institute, Inc.

Compressed Gas Association. 1984. 1995. *Anhydrous Ammonia.* Pamphlet G-2, 8th ed. Chantilly, Va.: Compressed Gas Association.

Connell, G.F. 1996. *The Chlorination/Chloramination Handbook.* Denver, Colo.: AWWA.

LaRoche Industries. 1989. *Ammonia Technical Data Manual.* Duluth, Ga.: LaRoche Industries.

———. 1997. *Aqua Ammonia Technical Data Manual.* Available at: www.airgasspecialtyproducts.com/default.aspx?PID=1778.

———. 2002. Material Safety Data Sheet No. 4002, Ammonium Hydroxide (9/16/02). Available at: http://www.airgasspecialtyproducts.com/UserDyn/laroche/4002MSDS-R-09-16-02.pdf.

———. 2003. *Anhydrous Ammonia Safety.* Duluth, Ga.: LaRoche Industries.

Wallace & Tiernan. 1991. *Ammonia Handling Manual.* Vineland, N.J.: Wallace & Tiernan.

Washington State Legislature. 2001. Storage and Handling of Anhydrous Ammonia. Chapter 296-24 WAC, Part F-2. General Safety and Health Standards. Available at: www.lni.wa.gov/wisha/rules/generalsafety/PDFs/24-Part-f2.pdf.

White, G.C. 1999. *Handbook of Chlorination and Alternative Disinfectants,* 4th ed. New York: John Wiley & Sons.

Chapter 6

Chlorine/Chloramine Disinfection Strategies

MICROBIAL/DISINFECTION BY-PRODUCTS DILEMMA

The purpose of disinfection is to destroy or inactivate pathogenic microorganisms to reduce the risk of exposure to waterborne diseases. Chlorine has been the primary disinfectant used for this purpose since its introduction in the early 1900s (AWWA 1973). Over most of that period the treatment strategy was to clarify the water to remove particles and use chlorine as the disinfectant. Chlorine was generally used in the lowest amount that would achieve the disinfection goal (due primarily to cost and taste considerations). This strategy has been highly effective in reducing waterborne disease.

Disinfection strategies have become much more complicated since trihalomethanes (THMs) were discovered in drinking water in the early 1970s (USEPA 1991). These compounds were found to be possible carcinogens and their formation is directly associated with the use of chlorine. This discovery has led to a restriction on the amount of chlorine used, limits on the concentration of THMs, and goals for the removal of organic precursors that lead to the formation of these compounds.

Several regulations in the United States are directed specifically at the use of disinfectants and the formation of disinfection by-products (DBPs). The interrelationship of these regulations has led to a dilemma: *Ensuring that drinking water is free from pathogens AND reducing the risk from potentially harmful disinfection by-products.* Water quality professionals are challenged to simultaneously satisfy both of these seemingly conflicting objectives. This chapter presents strategies that are being used to achieve the balance needed to satisfy these goals.

US Regulatory Framework

A progression of US regulations has responded to microbial disinfection needs and the growing concern about health consequences of DBPs. The interrelationships and conflicts between the current and proposed requirements are complex. It is important, however, to have an understanding of these issues as a utility develops its disinfection strategy (AWWA and ASCE 2005).

The US Environmental Protection Agency (USEPA) has developed a number of rules under the Safe Drinking Water Act (SDWA) and amendments that define the requirements for specific contaminants and processes. A number of these rules address microbial/disinfection requirements: Surface Water Treatment Rule, Total Coliform Rule, Interim Enhanced Surface Water Treatment Rule, Long Term 1 Enhanced Surface Water Treatment Rule, Long Term 2 Enhanced Surface Water Treatment Rule, and the Groundwater Rule. Several other rules have been developed to address DBP limitations: Total Trihalomethane Rule, Stage 1 Disinfectants and Disinfection By-Products Rule, and the Stage 2 Disinfectants and Disinfection By-Products Rule. The Lead and Copper Rule, although it is designed to address the levels of these contaminants, has affected disinfection and DBPs due to its pH control requirements. Brief descriptions of these rules are given in the following sections along with how they may affect chlorine/chloramines disinfection strategies. Regulations are modified periodically. Therefore, the information presented (although accurate at the time of preparation) may misrepresent the current status of the regulatory requirements. For current information on regulations, contact applicable regulatory agencies.

Surface Water Treatment Rule requirements. On June 29, 1989, the USEPA enacted the Surface Water Treatment Rule (SWTR). This regulation applies to every public water system in the United States that uses surface water as a source. It also applies to systems that use groundwaters that are vulnerable to intrusion of surface water; these systems are labeled as having "groundwater under the direct influence of surface water."

The purpose of the regulation is to protect the public from waterborne diseases that are most commonly transmitted by contamination of surface water (USEPA 1991). Because it is difficult to monitor for particular microorganisms, including *Giardia lamblia* and viruses, the SWTR emphasizes treatment techniques as the condition for compliance, rather than establishing maximum contaminant levels (MCLs) for specific microorganisms and viruses.

Because of the variances in water quality, local conditions, and methods of treatment, the rule does not prescribe a particular method of treatment; instead, it offers several alternatives. Any of these can then be used by a water system to meet the overall goal, which is removal or inactivation of disease-causing organisms. By allowing each water system to choose the best method of treatment for its situation, the rule makes it possible to protect public health and still provide safe water at the least possible cost. However, to ensure that water quality goals are met, the rule contains many operational and monitoring requirements.

Giardia cysts and viruses are among the most resistant waterborne pathogens. The SWTR assumes that water systems that attain adequate removal and/or inactivation of these organisms will provide adequate protection from other waterborne disease-causing organisms.

Most systems with a surface water source must use sedimentation and filtration to ensure adequate removal of microorganisms. Only systems with extremely clean source water may be allowed to operate without filtration, and they may do so only under very stringent operating and monitoring conditions. All surface water systems, whether they must provide filtration or not, must practice disinfection under very specific conditions.

The effectiveness of chlorination for inactivating *Giardia* cysts and viruses depends on the following:

- chlorine residual concentration (C),

- time the water is in contact with chlorine (T),
- water temperature, and
- pH of the water.

The residual concentration (C) of a disinfectant, in milligrams per liter (part per million) multiplied by the contact time (T) in min, equals the CT value. The CT values required by various disinfectants to guarantee the necessary reduction in microorganisms are given in appendix B. Discussions of determining the contact time T, calculating CT, and improving the contact time are presented later in this chapter.

Some of the principal disinfection requirements of the SWTR are as follows:

- Each system's disinfection treatment process must be sufficient to ensure that the total treatment process achieves at least 99.9 percent (3-log) inactivation and/or removal of *Giardia* cysts, and 99.99 percent (4-log) inactivation and/or removal of viruses.
- The disinfectant residual of water entering the distribution system must be monitored continuously by systems serving a population of more than 3,300. Systems serving less than 3,300 may take grab samples.
- The disinfectant residual in water entering the distribution system must not be less than 0.2 mg/L for more than 4 hr during periods when a system is serving water to the public. If it falls below this level, the system must notify the state.
- The disinfectant residual in the distribution system must be measured at the same points and at the same time that total coliforms are sampled. The disinfectant residual in these samples cannot be undetectable in more than 5 percent of the samples each month for any two consecutive months that water is served to the public.

Total Coliform Rule. The Total Coliform Rule (promulgated in 1989) establishes sampling and MCL requirements for coliform bacteria (AWWA and ASCE 1997). Compliance is based on the number or percentage of samples testing positive during a given month. If any sample is determined to be coliform positive, three repeat samples must be collected and analyzed within 24 hr, one each upstream and downstream of the original site. Additional sets of three samples are to be taken until all three are negative, or the monthly MCL is met.

- Whenever the MCL is exceeded, the primacy agency must be notified within 24 hr, and the supplier must prepare a report identifying the cause of the problem. The Rule requires proper public notification, using standard language as published in the regulation.

Lead and Copper Rule. The Lead and Copper Rule (promulgated in 1991) seeks to minimize lead and copper at users' taps. The Rule establishes action levels for lead (0.015 mg/L) and copper (1.30 mg/L) for the 90th percentile of the samples measured at customer taps. Monitoring for a variety of water quality parameters is required. In addition to monitoring, all large systems are required to conduct corrosion studies to determine optimal lead and copper corrosion control strategies.

If the action triggers are exceeded, the system is required to evaluate several approaches: public education, source water treatment, corrosion control practices, and possibly lead pipe replacement. Corrosion control can include pH/alkalinity adjustment, corrosion inhibitor addition, and calcium adjustment. A number of water quality parameters must be monitored to comply with this rule (e.g., lead, copper, pH, alkalinity, and calcium).

This rule can affect disinfection strategies because some of the control measures for lead and copper involve water chemistry adjustments (specifically pH control). These adjustments can affect the formation of DBPs and disinfection effectiveness.

Therefore, corrosion control measures employed to comply with the Lead and Copper Rule must also be considered in the selection of an overall disinfection strategy.

Trihalomethane regulations. Surface waters and groundwaters contain varying levels of organic compounds. These compounds generally come from natural sources, and they are primarily humic and fulvic acids. Typical examples of water with high levels of organic compounds are runoff water from forested land and water from lakes with high algae levels.

When chlorine is added to water containing these organic compounds, a reaction takes place to form chloro-organic compounds known as THMs. These compounds, the most common of which is chloroform, are considered potential cancer-causing substances. The USEPA set an MCL of 0.08 mg/L (effective Jan. 1, 2002; initially 0.1 mg/L in 1979) for total THMs in drinking water (AWWA 1999).

It is possible to remove THMs from treated water by using activated carbon, but this is an expensive solution. A preferred solution is to prevent the THMs from forming in the first place. One method to avoid THM formation is to use a disinfectant other than chlorine for the initial dose to the untreated water. Some other disinfectants will not form THMs; however, they may have other disadvantages and may form other DBPs. Many systems find that excessive THM formation is only a seasonal problem, such as during spring and fall runoff periods. In this case, special operating procedures may be necessary only during those periods.

Stage 1 Disinfectants and Disinfection By-Products Rule. The Stage 1 Disinfectants and Disinfection By-Products Rule applies to community water systems and nontransient, noncommunity systems, including those serving fewer than 10,000 people, that add a disinfectant to the drinking water during any part of the treatment process (USEPA 2001a).

The Rule includes the following key provisions:

- Maximum residual disinfectant level goals (MRDLGs) for chlorine (4 mg/L), chloramines (4 mg/L), and chlorine dioxide (0.8 mg/L).
- Maximum contaminant level goals (MCLGs) for four THMs (chloroform [zero], bromodichloromethane [zero], dibromochloromethane [0.06 mg/L], and bromoform [zero]), two haloacetic acids (HAAs) (dichloroacetic acid [zero] and trichloroacetic acid [0.3 mg/L]), bromate (zero), and chlorite (0.8 mg/L). *USEPA subsequently removed the zero MCLG for chloroform from its National Primary Drinking Water Regulations, effective May 30, 2000, in accordance with an order of the US Court of Appeals for the District of Columbia Circuit.*
- Maximum residual disinfectant levels (MRDLs) for three disinfectants (chlorine [4.0 mg/L], chloramines [4.0 mg/L], and chlorine dioxide [0.8 mg/L]).
- MCLs for total THMs (0.080 mg/L, which is a sum of the four listed earlier), haloacetic acids (HAA5) (0.060 mg/L, which is a sum of the two listed earlier plus monochloroacetic acid and mono- and dibromoacetic acids), and two inorganic DBPs (chlorite [1.0 mg/L] and bromate [0.010 mg/L]).
- A treatment technique for removal of DBP precursor material (USEPA 1999a).

The terms *MRDLG* and *MRDL*, which are not included in the SDWA, were created during the negotiations to distinguish disinfectants (because of their beneficial use) from contaminants. The final rule includes monitoring, reporting, and public notification requirements for these compounds. This final rule also describes the best available technology (BAT) upon which the MRDLs and MCLs are based.

Interim Enhanced Surface Water Treatment Rule. The Interim Enhanced Surface Water Treatment Rule applies to systems using surface water, or groundwater under the direct influence of surface water, that serve 10,000 or more persons (USEPA

2001b). The rule also includes provisions for states to conduct sanitary surveys for surface water systems regardless of system size. The rule builds upon the treatment technique requirements of the SWTR with the following key additions and modifications:

- MCLG of zero for *Cryptosporidium*.
- 2-log *Cryptosporidium* removal requirements for systems that filter.
- Combined filtered water effluent turbidity must be less than or equal to 0.3 ntu in at least 95 percent of the measurements taken each month, with measurements taken every 4 hr of operation.
- Combined filtered water effluent turbidity must not exceed 1.0 ntu at any time with measurements taken in 4-hr intervals.
- Special requirements for turbidity at slow sand and diatomaceous earth filtration plants.
- Individual filter effluents must be monitored continuously for turbidity.
- Any individual filter with a turbidity level greater than 1.0 ntu, based on two consecutive measurements 15 min apart, must be reported to the governing agency.
- Any individual filter with a turbidity level greater than 0.5 ntu at the end of the first 4 hr of filter operation, based on two consecutive measurements 15 min apart, must be reported to the governing agency.
- Disinfection profiles must be prepared by systems with total THM or HAA5 annual distribution system levels of 0.064 mg/L or 0.048 mg/L or higher (USEPA 1999b). The disinfection profiles will consist of daily *Giardia lamblia* log inactivation over a period of 1 to 3 years. These will be used to establish benchmarks for microbial protection to ensure that there are no significant reductions as systems modify disinfection practices to meet the Stage 1 DBP Rule.
- Systems using groundwater under the direct influence of surface water are now subject to the new rules dealing with *Cryptosporidium*.
- Inclusion of *Cryptosporidium* in the watershed control requirements for unfiltered public water systems.
- Requirements for covers on new finished water reservoirs.
- Sanitary surveys, conducted by states, for all surface water systems regardless of size.

The Interim Enhanced Surface Water Treatment Rule, with tightened turbidity performance criteria and required individual filter monitoring, is designed to optimize treatment reliability and to enhance physical removal efficiencies to minimize the *Cryptosporidium* levels in finished water. In addition, the rule includes disinfection benchmark provisions to ensure continued levels of microbial protection while facilities take the necessary steps to comply with new DBP standards.

Long Term 1 Enhanced Surface Water Treatment Rule. While the Stage 1 Disinfectants and Disinfection By-Products Rule applies to systems of all sizes, the Interim Enhanced Surface Water Treatment Rule only applies to systems serving 10,000 or more people. The Long Term 1 Enhanced Surface Water Treatment Rule, promulgated in January 2002, strengthens microbial controls for small systems (i.e., those systems serving fewer than 10,000 people). The rule will also prevent significant increase in microbial risk where small systems take steps to implement the Stage 1 Disinfectants and Disinfection By-Products Rule.

The Rule generally duplicates the approaches in the Interim Enhanced Surface Water Treatment Rule for improved turbidity control, including individual filter monitoring and reporting. The Rule also addresses disinfection profiling and benchmarking.

Stage 2 Disinfectants and Disinfection By-Products Rule. The proposed Stage 2 Disinfectants and Disinfection By-Products Rule seeks to prevent the exposure of individual customers to levels of DBPs in amounts that exceed standards as measured on a system-wide basis (USEPA 2003a). The Rule requires that systems comply with standards for total THMs and HAA5 on a locational running on an annual-average basis rather than on a system-wide basis. DBP levels will be monitored at sites where concentrations are highest rather than long resident sites. Sampling will occur according to a fixed schedule that targets peak occurrence times rather than allowing avoidance of these times.

All systems shall conduct an initial distribution system evaluation (IDSE) to identify the best monitoring sites. The IDSE will examine DBP sampling sites over a 1-yr period. There are provisions for states to wave IDSE requirements. Groundwater systems are allowed compliance extensions of 4 to 7 years depending on size.

Long Term 2 Enhanced Surface Water Treatment Rule. Under the proposed Long Term 2 Enhanced Surface Water Treatment Rule, large and medium-size surface water (or groundwater under the direct influence of surface water) systems will conduct a 2-yr source water monitoring program to characterize the occurrence of *Cryptosporidium* (USEPA 2003b). The method and sample size are specified. The results will determine into which of four treatment technique compliance "bins" a system will be placed. The bins demand a treatment technique response depending on the occurrence level of *Cryptosporidium*. Systems may avoid monitoring if they have 2 yr of qualified data or they already provide the highest level of treatment (equivalent to bin 4).

Treatment requirements associated with the *Cryptosporidium* occurrence level bins can be selected from a "toolbox" of compliance approaches that quantify log-removal performance of various watershed control, alternative source, pretreatment, improved filtration, and improved disinfection measures.

Unfiltered systems must continue to meet avoidance criteria and also achieve 2-log inactivation of *Cryptosporidium* using a minimum of two disinfectants. Although they are not subject to the bin-classification and toolbox approach to compliance, they are required to conduct the initial 2-yr monitoring.

Ground Water Rule. Although groundwater has historically been thought to be free of microbial contamination, recent research indicates that some groundwaters are a source of waterborne disease. Most cases of waterborne disease are characterized by gastrointestinal symptoms (e.g., diarrhea, vomiting) that are frequently self-limiting in healthy individuals and rarely require medical treatment. However, these same symptoms are much more serious and can be fatal for persons in sensitive subpopulations (i.e., young children, the elderly, and persons with compromised immune systems). In addition, research indicates that some viral pathogens found in groundwater are linked to long-term health effects (e.g., adult onset diabetes, myocarditis).

Presently, only surface water systems and systems using groundwater under the direct influence of surface water are required to disinfect their water supplies. The 1996 amendments to the Safe Drinking Water Act require the USEPA to develop regulations that require disinfection of groundwater systems "as necessary" to protect the public health. The proposed Ground Water Rule (GWR) will specify when corrective action (including disinfection) is required to protect consumers who receive water from groundwater systems (USEPA 2002).

This rule applies to public groundwater systems (systems that have at least 15 service connections, or regularly serve at least 25 individuals daily at least 60 days out of the year). This rule also applies to any system that mixes surface and groundwater if the groundwater is added directly to the distribution system and provided to consumers without treatment. The GWR does not apply to privately owned wells.

Some key elements of the proposed Rule include the following

- system sanitary surveys conducted by the state and identification of significant deficiencies;
- hydrogeologic sensitivity assessments for systems that are not disinfected;
- source water microbial monitoring by systems that do not disinfect and draw from hydrogeologically sensitive aquifers or have detected fecal indicators within the system's distribution system;
- corrective action by any system with significant deficiencies or positive microbial samples indicating fecal contamination; and
- compliance monitoring for systems that disinfect to ensure that they reliably achieve 4-log (99.99 percent) inactivation or removal of viruses.

Mandatory chlorination state requirements. Many states currently require all community public water systems to practice chlorination. In addition, the USEPA is considering more restrictive mandatory chlorination regulations that will probably affect all public water systems. Under the new requirements, systems having a protected groundwater supply and a history of good compliance with microbiological standards will probably be allowed a variance to avoid chlorination, at the discretion of the state.

Summary of regulations and disinfection implications. The US drinking water regulations have evolved since the SDWA was passed in 1974 (Figure 6-1). Much of the focus of the regulations has been on disinfection to ensure the safety of drinking water. The Total Coliform Rule, the Surface Water Treatment Rule, the Interim Enhanced Surface Water Treatment Rule, the Long Term 1 Enhanced Surface Water Treatment Rule, the Long Term 2 Enhanced Surface Water Treatment Rule, and the Ground Water Disinfection Rule are examples of regulations that have specifically addressed microbiological water quality.

After the discovery of THMs in drinking water, and with knowledge of the possible health consequences of these compounds, regulations were developed to limit the formation of DBPs. The Total Trihalomethane Rule, the Stage 1 Disinfectants and Disinfection By-Products Rule, and the Stage 2 Disinfectants and Disinfection By-Products Rule are examples of regulations that have sought to minimize the formation of DBPs.

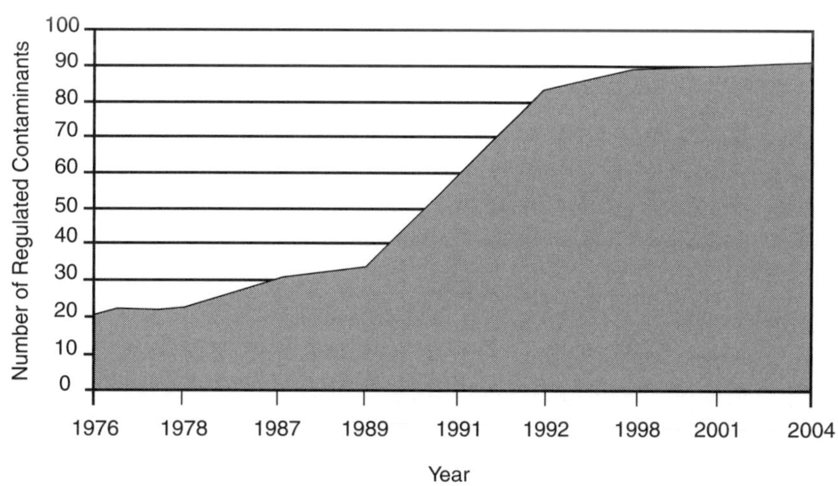

Source: www.epa.gov.

Figure 6-1 Number of regulated contaminants

Several regulations are primarily directed at other health issues or contaminants but have indirectly impacted either microbial control or the production of DBPs. Examples of these regulations include the Interim Primary Drinking Water Regulations and the Lead and Copper Rule. Interrelationships between these regulations that affect either microbial disinfection or the formation of DBPs are generally linked to the treatment conditions required to optimize particle removal, enhance organic precursor removal, and efficiently and effectively disinfect drinking water (with chlorine or chloramines in this case). The primary factors that affect disinfection are disinfectant type, disinfectant dose, the type of organism, contact time, pH, temperature, turbidity, and dissolved chemicals (organic and inorganic). The primary factors that affect DBP formation include pH, precursor concentration, temperature, and type of disinfectant. Many of the factors are identical and, thus, have led to conflicts when developing treatment strategies that can satisfy all of the regulatory requirements.

OPTIMIZING THE CHLORINATION/CHLORAMINATION DISINFECTION PROCESS

Several major operational adjustments can result in improvements in disinfection (maximize microbial protection) while minimizing DBP formation (USEPA 1999c). These adjustments should be employed in a treatment strategy that balances these sometimes competing goals.

Maximizing Microbial Protection

Disinfection effectiveness is improved by enhancing the contact time and optimizing the chemical environment. Employing these operational enhancements may also result in the production of more DBPs if the disinfection conditions are not carefully selected.

Enhancing contact time. The SWTR establishes requirements for inactivating *Giardia* and viruses under a variety of conditions. Each water system must treat to ensure that the total process achieves at least 99.9 percent (3-log) inactivation or removal of *Giardia* cysts and 99.99 percent (4-log) inactivation or removal of viruses. Source waters that are particularly vulnerable to microbial contamination may require greater log reductions.

Credit for physical removal of pathogenic organisms is generally given to properly operated filter plants as indicated in Table 6-1. Actual credit value must be granted by the appropriate regulatory agency. The remaining log reduction must be satisfied by the disinfection process.

Table 6-1 Surface Water Treatment Rule disinfection requirements

Process	Log Removals	
	Giardia Cysts	Viruses
Minimum log removal inactivation required	3	4
Conventional treatment credit	2.5	2
Remainder for disinfection	0.5	2
Direct filtration credit	2	1
Remainder for disinfection	1	3

The remaining log removal or inactivation shown in the table is achieved by a combination of disinfectant (chlorine) concentration (C) in milligrams per liter multiplied by the contact time (T) in minutes. Tables of CT values for differing pH, temperature, and log removals are given in appendix B. The contact time used in the calculation is the time it takes water to move from the disinfectant application point to the point at which the residual is measured.

Under most conditions, water does not move through reservoirs, tanks, and basins in a uniform manner. It is this factor that must be managed to efficiently use the disinfectant and minimize the need for unnecessarily large tanks. In general, tanks with a large length-to-width ratio (Figure 6-2) and with good inlet and outlet baffling minimize short-circuiting and provide the most uniform flow (Crozes et al. 1999) (Figure 6-3).

The contact time used to calculate the CT is the detention time at which 90 percent of the water passing through the reservoir is retained within the reservoir. That is, the time for 10 percent of the water to pass through the reservoir (T_{10}).

Tracer studies may be conducted to determine the actual T_{10} for a basin under a number of flow conditions. The method of conducting tracer studies is outlined in the *Guidance Manual for Compliance with the Filtration and Disinfection Requirements for Public Water Systems Using Surface Water Sources* (USEPA 1991).

Under most conditions, regulatory agencies allow the contact time for a reservoir to be determined by an approximation. The method involves multiplying the theoretical contact time (plug flow) by a factor that takes into account the baffling conditions of the basin. Examples of poor, average, and superior baffling conditions are shown in Figures 6-4, 6-5, and 6-6. Table 6-2 summarizes the baffling conditions and the proportion of the T_{10} that is allowed.

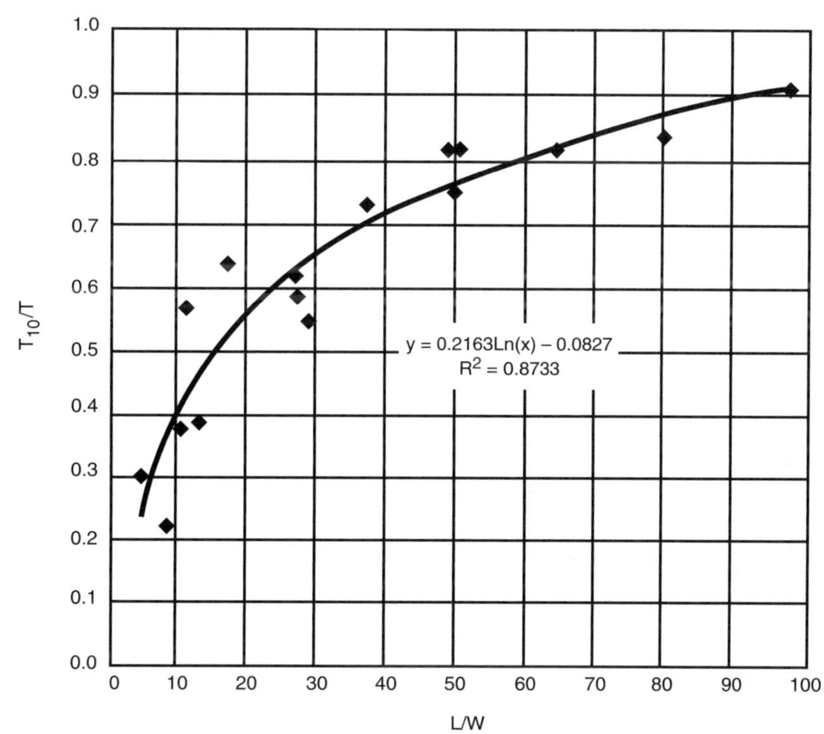

Source: Crozes et al., 1999.

Figure 6-2 Impact of L/W on T_{10}/T ratio

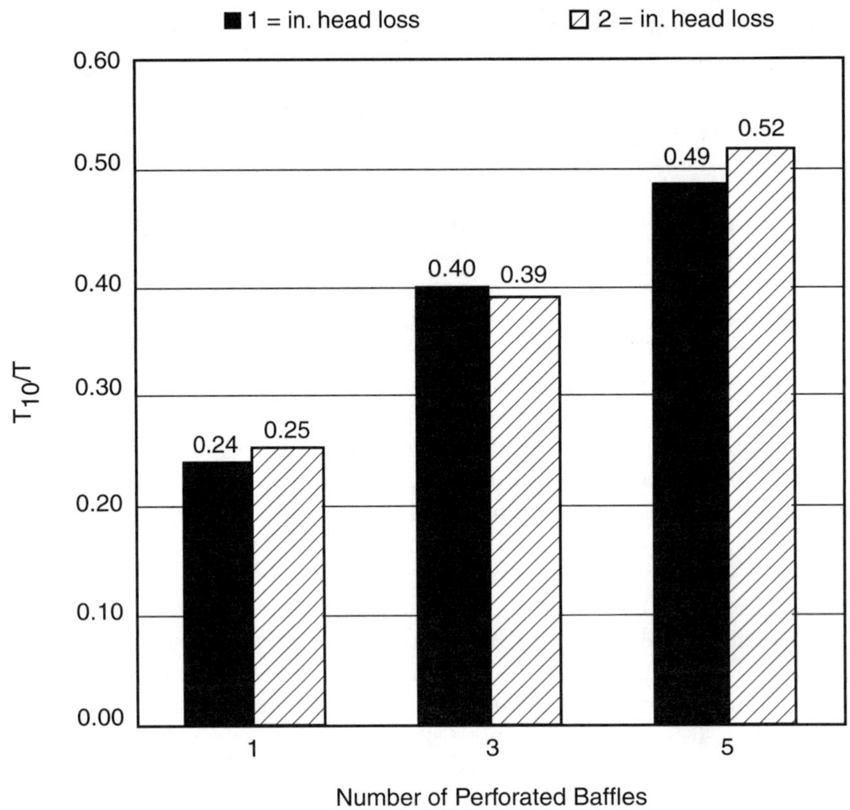

Source: Crozes et al. 1999.

Figure 6-3 Effect of perforated baffles on the T_{10}/T ratio of circular clearwells

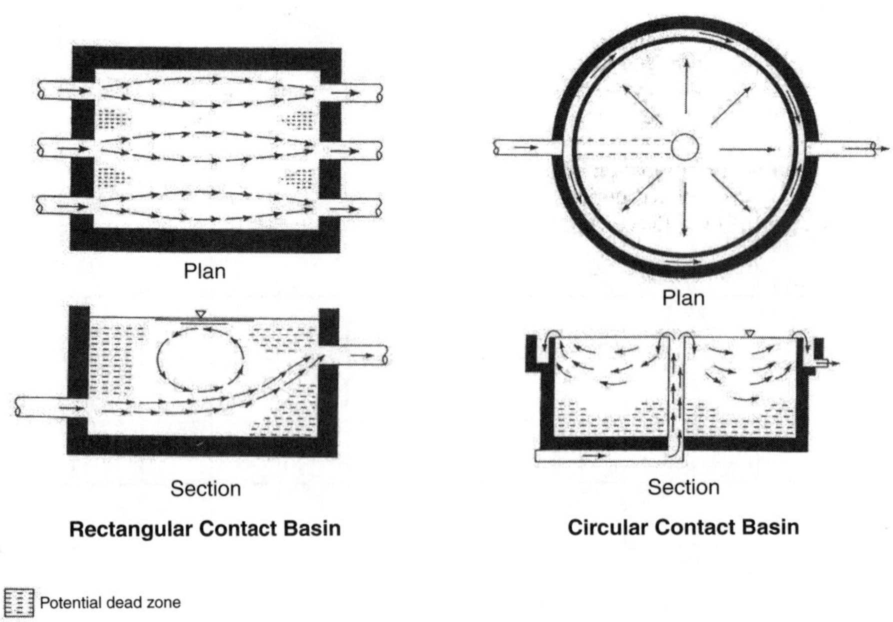

Source: AWWA and ASCE, 1997.

Figure 6-4 Examples of poor baffling conditions in basins

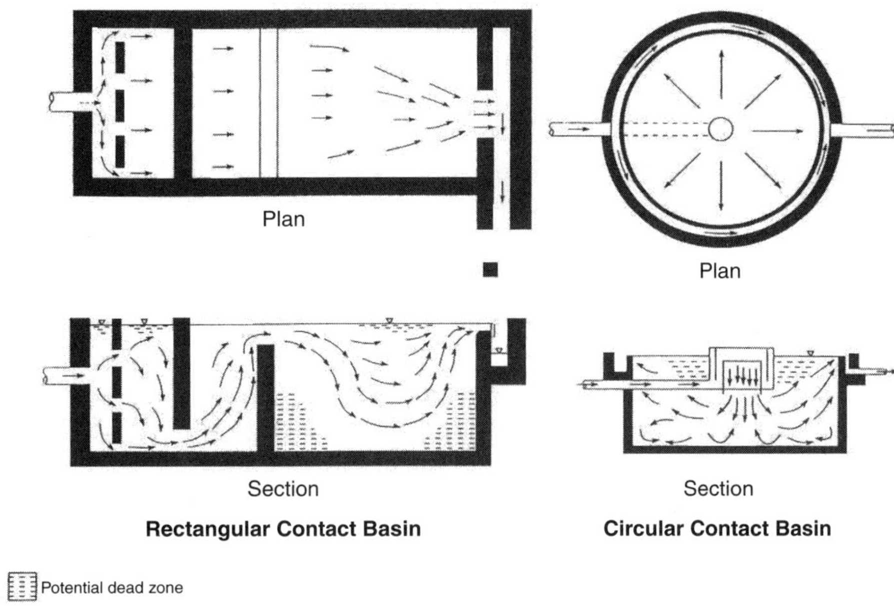

Source: AWWA and ASCE, 1997.

Figure 6-5 Examples of average baffling conditions in basins

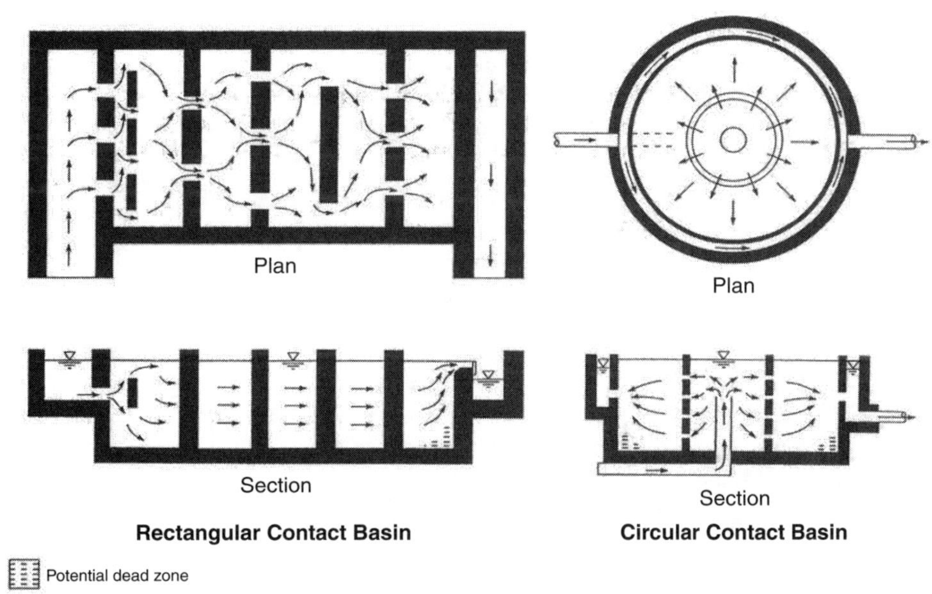

Source: AWWA and ASCE, 1997.

Figure 6-6 Examples of superior baffling conditions in basins

Table 6-2 Baffling classifications

Baffling Conditions	T_{10}/T	Description
Unbaffled	0.1	No baffling, agitated basin, very low length-to-width ratio, high inlet and outlet velocities
Poor	0.3	Single or multiple unbaffled inlets and outlets, no baffles
Average	0.5	Baffled inlet or outlet with some intrabasin baffles
Superior	0.7	Perforated inlet baffle, serpentine or perforated intrabasin baffles, outlet weir or perforated launders, filters
Perfect (plug flow)	1.0	Very high length-to-width ratio, perforated inlet, outlet, and intrabasin baffles, pipelines

Disinfectant (chlorine or chloramine) may be added at more than one location in a treatment plant and the CTs combined for compliance purposes. This approach may provide benefits by reducing DBP formation by applying most of the chlorine following precursor removal and including a chlorination step with modest dosages prior to filtration to enhance the performance of that process. An example is provided in the following section.

Example disinfection CT calculation. A 2-mgd direct filtration treatment plant requires 3-log *Giardia* cyst removal/inactivation. The plant is well operated and qualifies for a 2-log credit. *Giardia* cyst inactivation is the controlling parameter (virus requirements are more easily met in this case) and, therefore, disinfection must provide a 1-log inactivation for *Giardia*. Chlorine is added to the clearwell (80,000 gal with superior baffling). A 1-mile, 16-in. transmission main transports the water to the first customer. The plant is running at peak hourly flow of 1.5 mgd. The chlorine residual in the clearwell is 1.0 mg/L and 0.6 mg/L in the pipeline (pH is 7.5 and temperature is 5°C [41°F] in both locations).

To calculate the CT for the clearwell:

Contact time = 80,000 gal/1,500,000 gal/day × 1,440 min/day × 0.7 (baffle factor) = 54 min

Concentration = 1 mg/L

Required CT (from appendix B) for 3-log removal = 179 mg/L-min

$$\text{Fraction of required CT} = \frac{54 \text{ mg/L} - \text{min}}{179 \text{ mg/L} - \text{min}} = 0.30$$

To calculate the CT for the pipeline:

Contact time = 5,280 ft × ([16 in./2] × [1 ft/12 in.])2 × 3.14 × 7.48 gal/cu ft/1,500,000 gal/day × 1,440 min/day × 1 (baffle factor) = 53 min

Concentration = 0.6 mg/L

Required CT (from appendix B) for 3-log removal = 171 mg/L-min

$$\text{Fraction of required CT} = \frac{53 \text{ min} \times 0.6 \text{ mg/L}}{171 \text{ mg/L} - \text{min}} = 0.19$$

Total fraction of 3-log CT is 0.49 or log *Giardia* inactivation = 3 × 0.49 = 1.47.

1-log *Giardia* inactivation was required by disinfection so this plant meets the treatment requirement.

Adjusting pH for improved disinfection. Many factors such as temperature, time, and type of organism affect the efficacy and speed of the disinfection process, but pH is the most important factor for chlorination. The influence of pH on chlorination and chloramination is discussed in detail in chapter 3 (and in the CT discussion earlier in this chapter). In order to achieve compliance with the CT requirements, some systems adjust pH so that disinfection is enhanced. For chlorine, the CT requirements (for a fixed temperature) are less as pH is reduced (see appendix B); therefore, compliance may be achieved at the same residual in less time. This procedure sometimes allows a treatment plant to achieve compliance with the regulation without the need to add contact basin capacity.

The importance of the relationship between pH and disinfection effectiveness is magnified when considering the requirements of the Lead and Copper Rule (described earlier in this chapter). In many cases, to comply with the corrosion control requirements of the rule, the pH of the treated water must be increased. This procedure may achieve the lower lead and copper values at customer taps that is the focus of the rule (there are a number of other treatment practices that may also be satisfactory). However, raising the pH makes it necessary to provide longer contact time or to increase the chlorine residual to achieve the same disinfection credit. Given these facts, treatment plants often employ a strategy that applies chlorine prior to increasing the pH (at a more advantageous pH) to gain the maximum CT credit. The pH (for corrosion control) is then increased just as the water leaves the plant and after CT compliance has been achieved.

The effectiveness of chloramines is not influenced by pH in the same way as free chlorine. Disinfection credit (CT) for these compounds is not changed regardless of pH (within the 6–9 pH range). Despite the lack of regulatory disinfection credit adjusted for pH, there are important shifts in chloramine species that are influenced by pH. At lower pH, the percentage of dichloramine increases (at pH 6, approximately 20 percent of the total chlorine residual in dichloramine). This can be important even though dichloramine may be a more powerful disinfectant, and the odor threshold is much lower than monochloramine (approximately one-fourth). This may result in odor complaints as the pH is lowered.

The disinfection effectiveness of chloramines (primarily monochloramine at normal, higher pH values), however, should not be discounted. Studies have shown that monochloramines may match the effectiveness of free chlorine (inactivation of coliform organisms) when contact times are more than 45 min. Additionally, chloramines have shown superior performance when facing established biofilms. These results have led to the wide use of chloramines as residual disinfectants in distribution systems.

Although this section is focused primarily on the effect of pH of disinfection effectiveness, it also impacts the formation of DBPs. The formation of many DBPs (HAAs, haloketones, trichloropropanone, dihaloacetonitrile, and chloral hydrate) decreases with increasing pH. Total THMs, however, are generally increased with increasing pH.

Minimizing Disinfection By-Product Formation

There are several practices that can be used to help reduce the formation of DBPs, including optimizing organic precursor removal, altering the location of disinfectant addition, and using chloramines as a residual disinfectant in the distribution system. A knowledge of the role of pH in the formation of DBPs (as indicated earlier) can also assist in their control.

Optimizing organic precursor removal. One strategy to reduce the formation of DBPs is to remove organic precursors prior to the addition of chlorine. Raw water can include DBP precursors in dissolved and particulate forms. In conventional treatment, dissolved precursors must be converted to particulate form for subsequent removal during settling and filtration (USEPA 1999a). The potential THM formation may be lessened by conventional coagulation and settling in certain waters, which may argue for moving the point of chlorine application downstream of coagulation and settling (and even filtration) to control the formation of HAAs and total THMs. Systems may lower the DBP formation potential of water prior to disinfection by removing precursors with bank filtration, an infiltration gallery, enhanced coagulation, lime softening, granular activated carbon adsorption, powdered activated carbon, or membrane filtration.

Aluminum (alum) and iron (ferric) salts both have the ability to reduce pH, which improves natural organic matter (NOM) removal. For alum, the optimal pH for NOM removal is in the range of 5.5 to 6.0 (enhanced coagulation). The addition of alum decreases pH and may allow the optimal pH range to be reached without acid addition. However, waters with very low or very high alkalinities may require the addition of a base or an acid to keep the pH in the optimal range. Systems must evaluate the best combination of coagulant, dosage, acid or base usage, coagulant/filter aids, etc.

Granular activated carbon (GAC) adsorption can be used to remove additional NOM. For most applications, empty bed contact times in the range of 5–30 min are required, with GAC regeneration frequencies on the order of 2 to 9 months. Changing the pH, or adding a disinfectant to the GAC bed, can result in specific reactions in which previously adsorbed compounds may be released into the treated water. Powdered activated carbon (PAC) can be used seasonally for precursor or total THM reduction. PAC is generally fed in a slurry into the treatment stream prior to filtration and usually before settling. Activated carbon (GAC or PAC) can react with chlorine and this reaction may interfere with the adsorption of organic compounds. Therefore, activated carbon is used prior to chlorine addition, and sometimes it is even used to *dechlorinate* the water.

Membrane filtration has been shown to be effective in removing DBP precursors in some instances. Ultrafiltration with a molecular weight cutoff of 100,000 daltons is ineffective for controlling DBP formation. Nanofiltration (molecular weight cutoff of 200–300 daltons), however, has removed precursors effectively. There are a number of significant limitations in the use of membranes including the disposal of the waste brine generated, fouling of the membranes, membrane replacement cost, and total system cost.

Altering the location of disinfectant addition. Disinfectants have commonly been applied at two points: where the raw water enters a treatment plant and again after treatment has been completed (Connell 1996) (Figure 6-7). A growing number of treatment plants have also found it advantageous to add disinfectants at other intermediate points in the treatment process.

In surface water treatment plants, common points of chlorine application include the raw water intake, flash mixer, filters, and clearwell. Other locations may be used from time to time depending on the plant process train. For example, in lime-softening plants, chlorine should not be added until the pH is returned to the 7–8 range; this is after the clarifiers (where calcium and magnesium salts have been removed), settling basins, and recarbonation process. Treatment with chlorine after the lime addition (under high pH conditions) could result in insufficient kill, would call for higher doses of chlorine, and may not meet the required CT value.

As illustrated in Figure 6-7, many surface water systems apply chlorine (or alternative disinfectants) at two points (Von Huben 1995). Source water chlorination (prechlorination) is performed to

- begin the process of killing and/or inactivating pathogenic organisms;
- minimize operational problems and tastes and odors that could be caused by biological growths on filters, pipes, and basins;
- oxidize hydrogen sulfide, iron, and manganese that may be in the raw water; and
- oxidize various organic substances in the raw water.

Historically, the most common point for prechlorinating surface water is the intake well or rapid-mix basin. Because the formation of THMs and other DBPs is a concern, many systems are moving the application point to later in the treatment process and/or feeding a different chemical oxidant as the water enters the treatment plant.

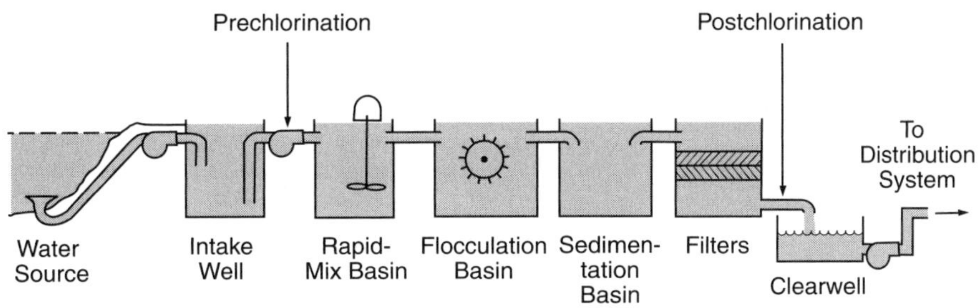

Figure 6-7 Common chlorination points in a conventional filtration plant

Optimizing chloramination. Chloramines may be used as a primary or secondary disinfectant. The major benefits include residual persistence to reach to the end of many distribution systems, effectiveness as a secondary disinfectant and the ability to penetrate biofilms in distribution systems, a tendency not to form THMs and other DBPs, and a minimization of chlorinous tastes and odors. Many utilities have turned to chloramines as the secondary (distribution system residual) disinfectant primarily to reduce the formation of DBPs. This application has proved effective.

There are, however, some important problems that may result from the use of chloramines. Operators must be aware of these potential consequences and institute procedures to minimize the impact. The major chloramination issues include the potential effects on special water uses such as kidney dialysis and fish rearing, possible effects on elastomeric materials used in distribution systems and plumbing fixtures, and vulnerability to the microbiological process known as nitrification.

Special water users should be notified that chloramines are being used and that their equipment or procedures should be modified to remove this chemical. Deterioration of elastomers by chloramines is enhanced by higher water temperatures. Operators in climates where high water temperatures are encountered should consider this potential issue when selecting the most suitable disinfectant.

Nitrification is a microbial process during which ammonia is sequentially oxidized to nitrite and nitrate. This may lead to rapid reduction in chloramine residual and increased growth of heterotrophic bacteria. There are many factors that can contribute to nitrification. (For a more complete discussion of this complex issue consult the references at the end of this chapter.) Most systems can control nitrification by the following:

- *Controlling the chlorine:ammonia-nitrogen ratio.* The chlorine:ammonia-nitrogen ratio should be maintained between 4.5:1 and 5:1 to reduce the concentration of free ammonia.
- Limiting the excess free ammonia (below 0.10 mg/L N).
- *Maintaining a set chloramine residual.* Systems that maintain a good chloramine residual everywhere in their system are less likely to encounter nitrification.
- *Maintaining pH and temperature.* The optimum pH range to minimize nitrification is 7.5–9.0 and the temperature is 50–59°F (10–15°C). Higher temperatures (77°F) (>25°C) have been associated with nitrification incidents.
- *Switching periodically to free chlorine.* Some systems find that switching to free chlorine periodically for short periods may help reduce the chloramine-resistant bacteria and reduce nitrification. However, switching to free chlorine may cause the system to exceed the DBP Rule, and the system may encounter biofilm sloughing.

- *Optimizing chemical addition and mixing* where the chemicals are added and in what order may influence the production of DBPs and the formation of desirable chloramine species (monochloramine instead of dichloramine). Mixing can help with the efficient formation of chloramines.
- *Monitoring ammonia, nitrite, and nitrate.* Careful surveillance of the distribution system (particularly in storage tanks and other areas where water may age) for free ammonia and the by-products of nitrification (nitrite and nitrate) may provide early warning signs of potential nitrification problems.
- *Attending to storage facility operation and maintenance.* Mixing may be needed to eliminate the development of problem areas within the storage facility. Operational strategies should minimize water age. Routinely clean and inspect all storage tanks and reservoirs.
- *Operating distribution systems to minimize nitrification* by decreasing water age, practicing systematic flushing, evaluating booster chloramination or chlorination, providing adequate corrosion control, and evaluating a periodic switch to free chlorine (may be required in some states).

DISTRIBUTION SYSTEM CHLORINATION (BOOSTER OR SECONDARY CHLORINATION)

It may be necessary to add chlorine to water after it has been sent to the distribution system. This addition, performed in the distribution system, is called *secondary* or *booster* chlorination. Extended distribution systems, large finished water reservoirs, or systems that provide only a free chlorine residual can easily lose some or all of the chlorine residual in the water. Growths in the distribution system may exhibit a chlorine demand and can exhaust much of the chlorine residual. This could leave the distribution system without any chlorine residual.

When this happens, the utility must take corrective action. One strategy is to provide additional points of chlorination at appropriate locations in the distribution system (Figure 6-8). Continuous monitoring of the chlorine residual in the distribution system or frequent sampling at specific locations in the system will alert operating personnel to a reduction in or absence of a residual. This sampling or monitoring can also be used to determine the location where the residual has been consumed. Booster stations may be located at any appropriate position in the system and can be automated to operate when necessary.

System operators must be aware that adding chlorine in the distribution system may result in an increase of DBPs. The Stage 2 Disinfectants and Disinfection By-Products Rule requires utilities to conduct an IDSE to identify the best DBP monitoring sites. The IDSE will examine DBP sampling sites over a year-long period. Changes in the location (or applied dosage) of chlorination booster stations may affect the monitoring site selection and the test results.

Supplementing a free chlorine residual with additional free chlorine is straight forward. However, when a system uses chloramines as the residual disinfectant in the distribution system, booster chlorination can become more complicated. Several studies have been conducted that examine this situation. The results have led to the development of site-specific operational practices that optimize this procedure.

When chlorine is added to water that has been previously chloraminated, there can be several outcomes. The chlorine can combine with the free ammonia remaining in the water to form chloramines again. The chlorine could react with the residual chloramine, possibly destroying the chloramine (according to the breakpoint curve shown in Figure 6-9) and resulting in a free chlorine residual. The chlorine may react with the chloramine and residual ammonia to form a mixture of chlorine species, some

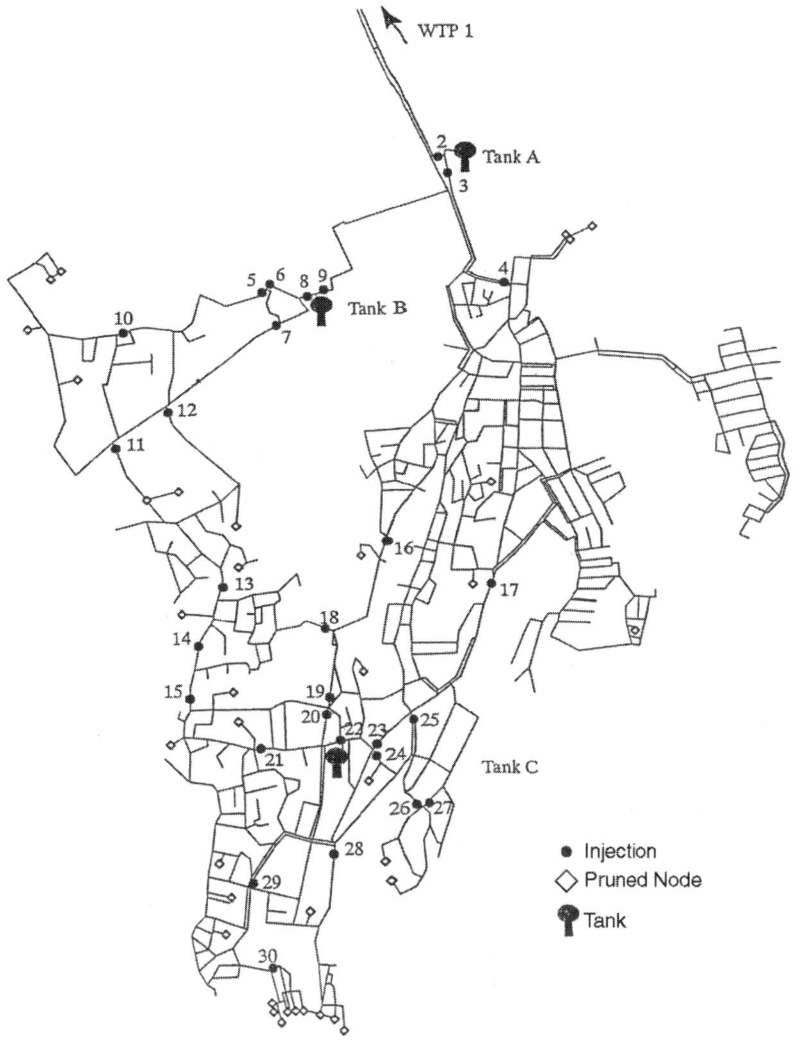

Figure 6-8 Network map with potential booster chlorination locations

of which could be less effective disinfectants or could cause odor complaints. Utilities have successfully employed several strategies for booster chlorination of chloraminated water, including the following:

- *Add free chlorine to combine with residual ammonia.* The procedure utilizes residual free ammonia remaining in the water as a consequence of chloramine decay. Chlorine is added in the ratio of 4.5:1 to 5:1 (chlorine:ammonia-nitrogen ratio) and a chloramine residual is reformed (Figure 6-10). This process has an added benefit since free ammonia is combined and, therefore, is not available to cause nitrification problems.
- *Add chlorine and ammonia.* Chloramine booster stations are rare, but they have been used in situations where there is not enough free ammonia remaining to reform sufficient chloramine concentrations for residual disinfection. Some stations use sodium hypochlorite and aqua ammonia to form the chloramine.

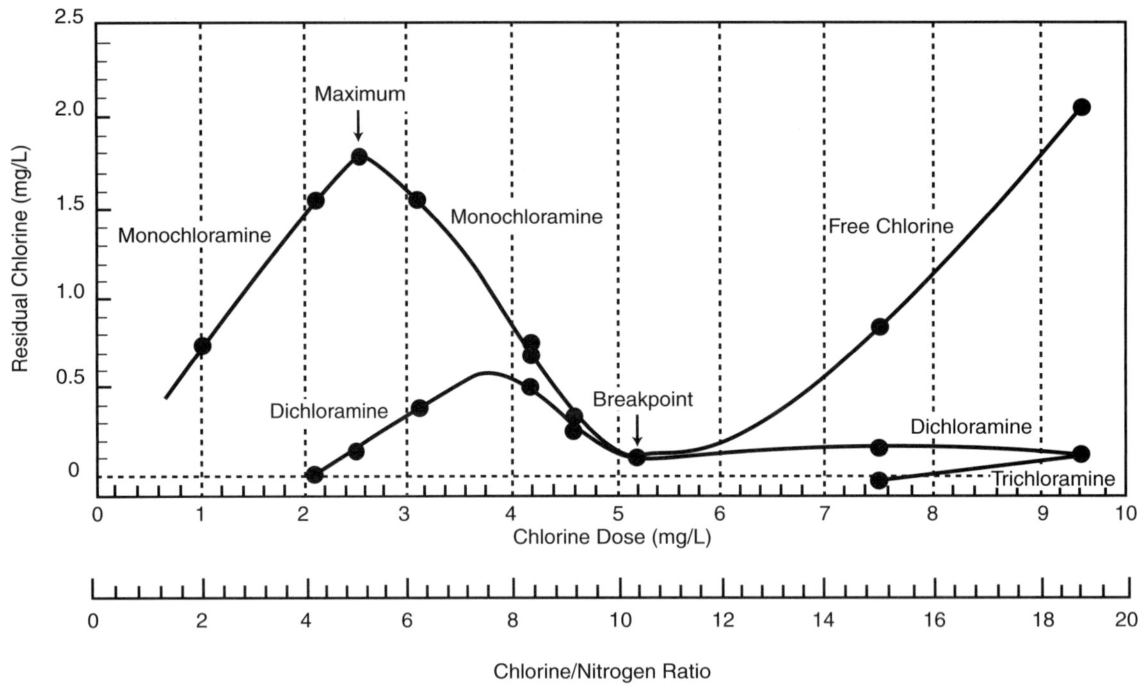

Figure 6-9 Example breakpoint chlorination curve

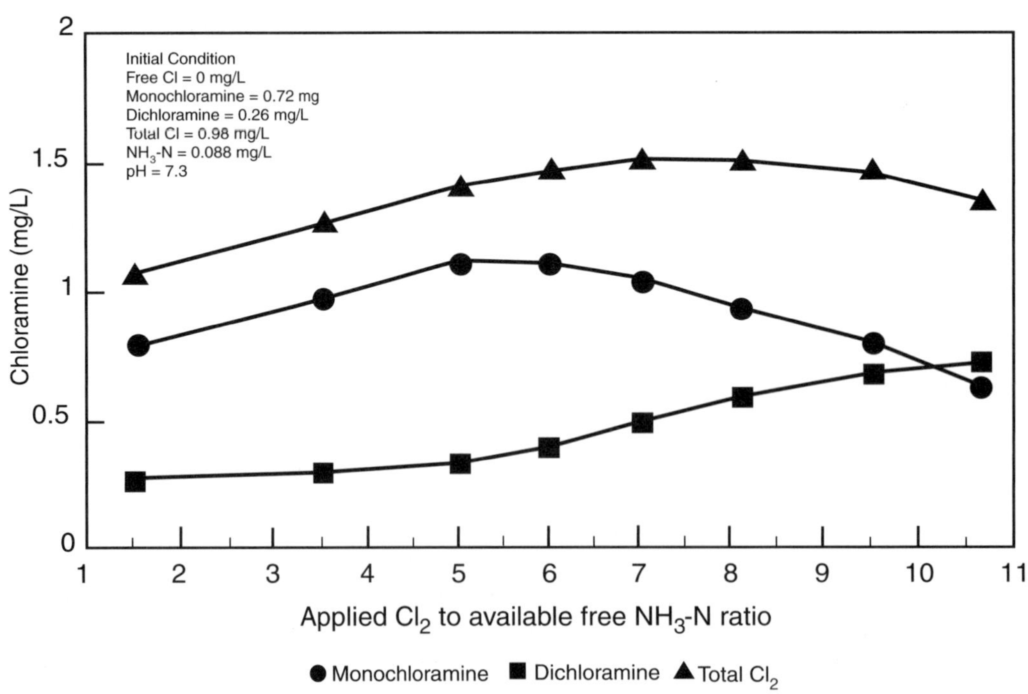

Figure 6-10 Booster chlorination breakpoint curve, 1 hr incubation

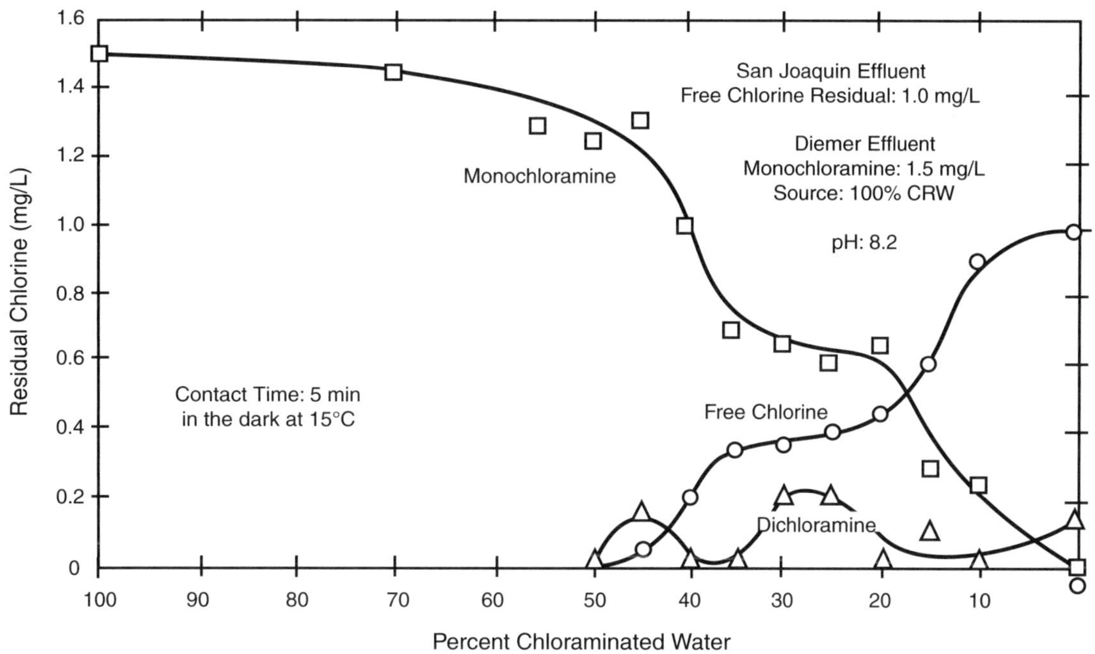

Figure 6-11 Blend-residual curve for San Joaquin Reservoir chlorinated effluent and diemer chlorinated effluent (5 min contact time)

- *Add chlorine to breakpoint.* This strategy is employed when a free chlorine residual is desired. Chlorine is added to water that contains residual ammonia and/or a small amount of chloramine. The amount added is sufficient to surpass the breakpoint (Figure 6-9) and achieve a free chlorine residual. This process eliminates any nitrification potential due to the presence of free ammonia. This procedure is used mostly where there is a separation between two systems or an area of one system can be isolated as long as THMs are not a problem in the area to be kept on free chlorine. One system or area can maintain a free chlorine residual and the other a chloramine residual.

Blending Chlorinated and Chloraminated Waters in Distribution Systems

Many water utilities buy water from neighboring utilities through common or interconnected pipelines (Connell 1996). This situation requires a special review of the residual desired by the purchaser and that available from the supplier. If both systems want a free chlorine residual, the addition of chlorine (appropriately controlled) to reach a desired residual is relatively simple.

Complexities develop when waters produced by the two systems do not have the same type of residual (free or combined) or level of residual and the two waters are to be blended (Figure 6-11). Several utilities have faced this issue and have devised strategies to provide adequate disinfectant residuals without adverse consequences.

When water containing chloramines is blended with water containing free chlorine, there are a number of possible outcomes, including the following:

- Chlorine may react with free ammonia (if present) to form chloramines.
- Chlorine may react with the chloramines, destroying the chloramines and resulting in a free chlorine residual (breakpoint chlorination).
- Chlorine and chloramines may react to form a number of chlorine species (e.g., free chlorine, monochloramine, dichloramine, trichloramine) in unpredictable proportions (Figure 6-11).

The concentration of chlorine and chloramines, the chlorine:ammonia-nitrogen ratio, the pH, the temperature, and the proportion of water containing each disinfectant all influence the potential outcome. Before considering blending of chloraminated and chlorinated waters, a utility should carefully examine the specifics of their situation. Studies should be conducted to thoroughly evaluate the consequences to ensure that customers are served with water that is protected with an acceptable disinfectant residual and there are no undesirable outcomes.

TREATMENT PLANT CHLORINATION/CHLORAMINATION STRATEGIES

Operators are faced with a myriad of regulations and the limitations of their treatment facilities but must still satisfy disinfection requirements and, at the same time, attempt to reduce the production of DBPs. To simplify this task, several general strategies are presented here. The reader should recognize that each situation is unique and the strategies presented here may not be the best for the specifics of their treatment plant. However, the general guidelines used to establish these example scenarios should be useful when considering the best treatment options for any plant.

General trends for chlorine/chloramine disinfection and THM and HAA control include the following:

- Chlorine disinfection is improved at lower pH and higher temperature.
- Chloramine disinfection is improved at higher temperature.
- THM formation is decreased at lower pH, chlorine residual, temperatures, and levels of organic precursors.
- HAA formation is decreased at higher pH, lower temperatures, and when organic precursors are reduced.
- Shorter contact time for chlorine, and to a lesser extent chloramines, with organic precursors results in lower THM and HAA concentrations.

Taking advantage of these trends, treatment plants can employ the following general strategies without jeopardizing disinfection (CT) effectiveness:

- reduce organic precursors before chlorine/chloramines are added,
- add chlorine/chloramines at a point where the pH is relatively low, and
- improve chlorine/chloramines contact time by improving mixing.

Conventional Treatment Plant Example

Figure 6-12 illustrates the traditional and modified chlorine application points in a conventional water treatment plant. Note that the primary disinfectant is commonly added at the rapid mix basin (for CT compliance). To reduce DBPs, that application point is now more commonly either just prior to filtration or at the clearwell. Clearwells are modified to improve the contact time (baffling provided) and satisfy CT requirements. Where oxidation prior to filtration is desirable, alternate oxidants such as potassium permanganate or chlorine dioxide may be used instead of chlorine. It is now becoming

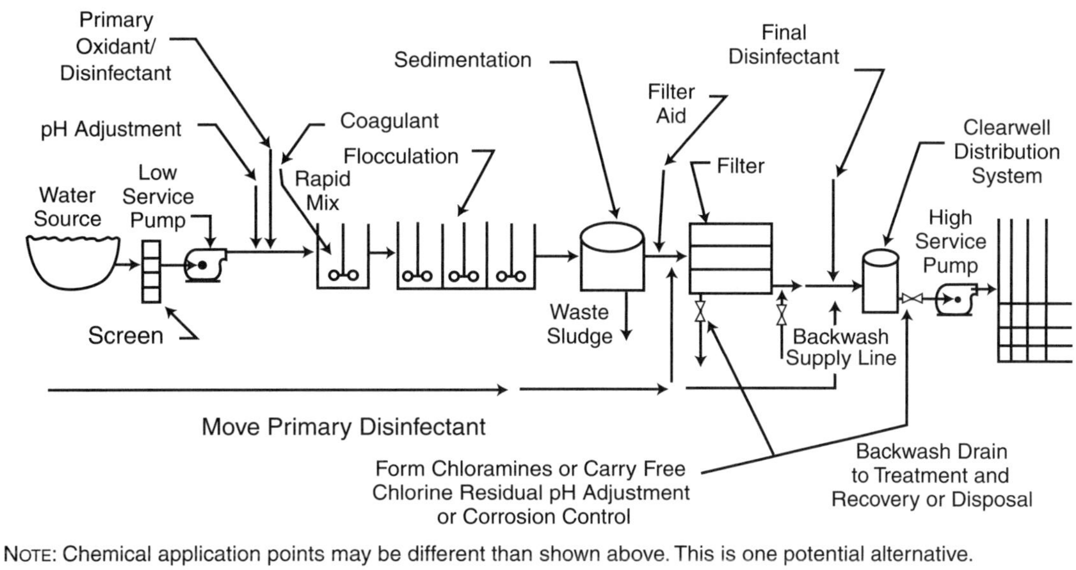

Source: AWWA and ASCE, 1999.

Figure 6-12 Conventional treatment, surface water

more typical for the residual disinfectant to be chloramines. They are commonly formed after the clearwell by adding the correct dosage of ammonia to combine with the free chlorine present (chlorine may need to be added as well to produce the correct residual level). Where pH adjustment or corrosion inhibitors must be used for corrosion control, these are typically added following the clearwell so that disinfection can take place under more advantageous conditions. It is possible that adequate contact time may be available to satisfy the CT with chloramines alone. In these cases, the chloramines may be added early in the treatment sequence to provide the necessary contact time.

Other Treatment System Examples

Two additional treatment system schematics are shown in Figures 6-13 and 6-14. In each case, the traditional points of chlorine/chloramines addition are shown along with the typical modified application points that are employed to maintain disinfection effectiveness while reducing DBPs. The primary strategy is to remove organic precursors prior to the addition of chlorine. Chloramines are sometimes used as the residual disinfectant. In that case, the CT requirement is satisfied by free chlorine followed by the formation of the chloramine residual. In some cases, adequate contact time may be available to meet the CT requirements with chloramines. To get the maximum contact time under these conditions, chloramines may be formed early in the plant treatment sequence and a chloramine residual carried through the entire plant.

Membrane treatment plants are similar in the major components but differ due to the array of membrane choices. Microfiltration and ultrafiltration membranes generally do not remove organic precursors. In these plants, addition of chlorine before and after the membranes may not greatly influence the production of DBPs. The location of chlorine addition may be dictated by the resistance of the membranes. Nanofiltration and reverse osmosis membranes may remove organic precursors and adding chlorine following the membranes should reduce the formation of by-products. Some of these membranes are resistant to chlorine or chloramines and may benefit from

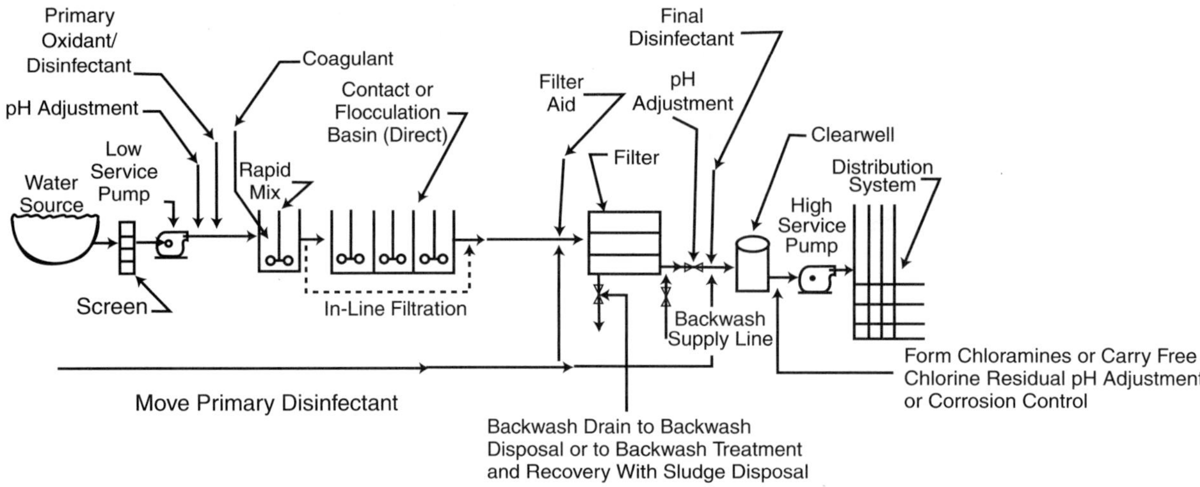

Source: AWWA and ASCE, 1999.

Figure 6-13 Direct and in-line filtration treatment, surface water

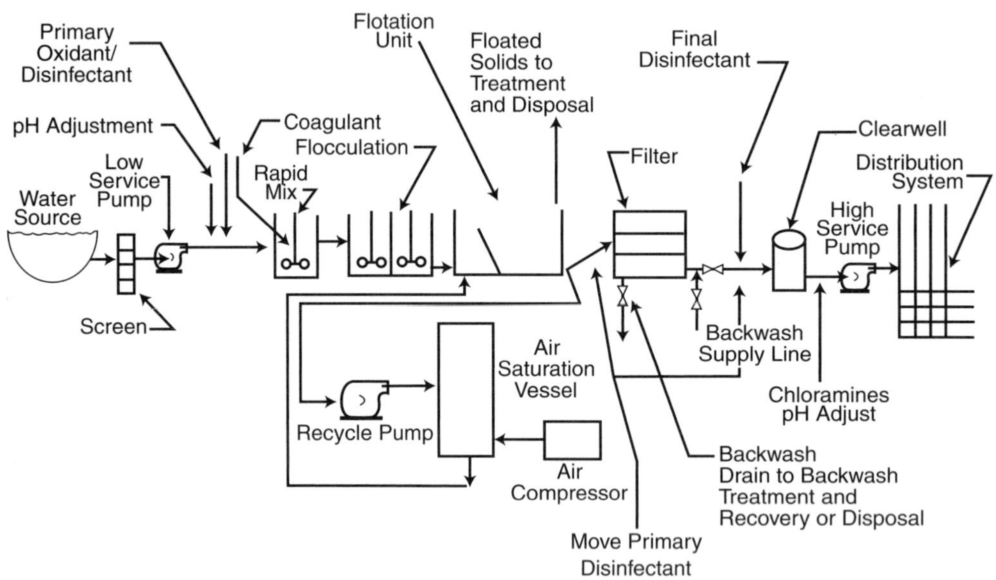

Source: AWWA and ASCE, 1999.

Figure 6-14 Dissolved air flotation/filtration treatment, surface water

their use prior to the membrane modules to reduce fouling. The formation of by-products may not be a problem with these systems since the contact time is so short.

Groundwater Example

Groundwater benefits from natural filtration (Connell 1996). As a result, the waters pumped from wells or obtained from artesian wells and springs are often low in concentrations of organic contaminants. There are a number of notable exceptions when dissolved organics in groundwater can contribute to the formation of chlorination

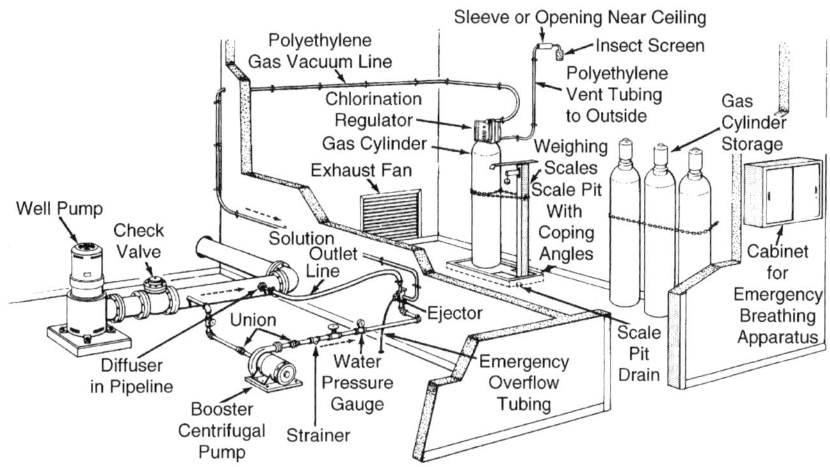

Courtesy of Severn Trent.

Figure 6-15 Typical chlorinator deep-well installation showing booster pump

by-products. Groundwater often does have materials dissolved in it or acquires materials during its passage through the subsurface filtration system. These materials are usually inorganic when of natural origin (e.g., not volatile organic chemical or synthetic organic chemical contamination), which, in the presence of chlorine, do not react to form THMs. Because groundwater generally contains very low concentrations of natural organic matter, THMs and HAAs are not significantly formed in the finished water.

Assuming a good quality source, chlorination is generally the only treatment employed in the treatment of well water. In these cases, chlorine is added to oxidize the unwanted inorganics (e.g., sulfide, iron, and manganese) and provide a residual disinfectant to the water as it enters the distribution system. Iron and manganese are not considered to be health risks but contribute to the formation of stains, bad taste, and iron bacteria growth in the distribution system.

Most water systems that use underground wells require only chlorination. Wells influenced directly by surface water infiltration, such as those shallow wells in close proximity to rivers, must satisfy the treatment technique requirements to meet the SWTR. A typical well chlorination system is illustrated in Figure 6-15.

In most well water systems, water is maintained under pressure from the well and may pass through a pressure filter system to remove oxidized inorganics before entering the distribution system. Because the chlorine demand in the water is often constant, the chlorination system need only be set manually for the correct dosage and activated automatically with the operation of the well pump. Sufficient CT and dosage must be provided to meet the required CT value prior to the first service connection and to meet state regulations regarding the desired residual in the distribution system.

Prechlorination for Zebra Mussel Control

Zebra mussels are indigenous to the fresh waters of Europe (Connell 1996). It is believed that, sometime during the 1980s, ships arriving in the Great Lakes from eastern European ports pumped bilge or ballast water containing zebra mussel larvae and/or adult mussels from the waters of their European sources. There have been many reported sightings in seemingly unrelated locations in the West, Southwest, East Coast, and Southeast. Figure 6-16 shows reported sighting locations as of 2004.

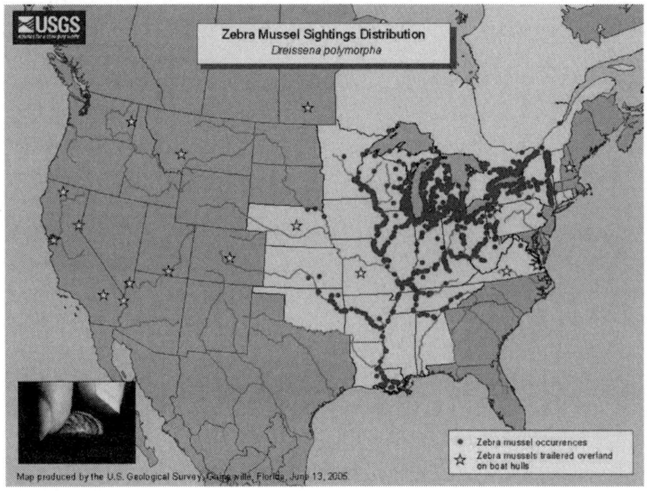

Source: *USGS.*

Figure 6-16 Zebra mussel sightings distribution, as of 2004

Zebra mussels create problems for water treatment plant operations by attaching themselves to the intake structure, piping, bar screens, and other components of the plant water system. Effective treatment of zebra mussels is achieved by the addition of chemical oxidants to the intake structure. One of the most effective oxidants is chlorine. Usually this treatment is accomplished by shock treatment using intermittent dosages of several parts per million of chlorine.

Shock treatment is the addition of higher than normal dosages of chlorine for shorter periods of time. This method results in lower total chlorine consumption and the ability to effect a more complete and rapid kill. THM formation consequences may need to be considered when using shock treatment. Because zebra mussels spawn during the times of the year when the water is warmer, shock treatment during the winter is generally not practiced. Dosages of 2–5 mg/L of chlorine for short time periods are often effective against zebra mussels. Shock dosages of 10 mg/L have also been employed for periods of 30 min and have been found to be effective.

It may be advantageous or necessary to use other forms of treatment, such as ozone, chlorine dioxide, or potassium permanganate, to kill zebra mussels. Although bromine compounds, such as bromine chloride and activated bromide salts, have been used in zebra mussel abatement, their use at water treatment intakes is not recommended. The presence of bromide enhances DBP formation when chlorine is added.

Each site has specific conditions that must be evaluated in regard to a zebra mussel problem. Operating personnel must analyze their local situation to determine the most practical and cost-effective means of resolution.

REFERENCES

AWWA (American Water Works Association). 1999. *Water Quality and Treatment,* 5th ed. New York: McGraw-Hill.

AWWA and ASCE (American Society of Civil Engineers). 2005. *Water Treatment Plant Design,* 3rd ed. New York: McGraw-Hill.

Connell, G.F. 1996. *The Chlorination / Chloramination Handbook.* Denver, Colo.: AWWA.

Crozes, G.F., J.P. Hagstrom, M.M. Clark, J. Ducoste, and C. Burns. 1999. *Improving Clearwell Design for CT Compliance.* Denver, Colo.: Awwa Research Foundation.

USEPA (US Environmental Protection Agency). 1991. *Guidance Manual for Compliance with the Filtration and Disinfection Requirements for Public Water Systems Using Surface Water Sources.* Washington, D.C.: USEPA.

———. 1999a. *Enhanced Coagulation and Enhanced Precipitative Softening Guidance Manual.* EPA-815-R-99-012. Washington, D.C.: USEPA.

———. 1999b. *Disinfection Profiling and Benchmarking Guidance Manual.* EPA-815-R-99-013. Washington, D.C.: USEPA.

———. 1999c. *Microbial and Disinfection Byproduct Rules Simultaneous Compliance Guidance Manual.* EPA-815-R-99-015. Washington, D.C.: USEPA.

———. 2001a. *Stage 1 Disinfectants and Disinfection Byproducts Rule: A Quick Reference Guide.* EPA 816-F-01-010. Washington, D.C.: USEPA.

———. 2001b. *Interim Enhanced Surface Water Treatment Rule: A Quick Reference Guide.* EPA 816-F-01-011. Washington, D.C.: USEPA.

———. 2002. Pre-proposal draft Ground Water rule language. Washington, D.C.: USEPA. [Online]. Available: http://www.epa.gov.

———. 2003a. *Draft Stage 2 DBPR Rule Implementation Guidance.* Washington, D.C.: USEPA. [Online]. Available: http://www.epa.gov.

———. 2003b. *Draft Long Term 2 Enhanced Surface Water Treatment Rule Implementation Guidance.* Washington, D.C.: USEPA. [Online]. Available: http://www.epa.gov.

Von Huben, H. 1995. *Water Treatment: Principles and Practices of Water Supply Operations,* 2nd ed. Denver, Colo.: AWWA.

ADDITIONAL SOURCES OF INFORMATION

Barrett, S.E., M.K. Davis, and M.J. McGuire. 1984. Blending Chloraminated Water With Chlorinated Water: Considerations for a Large Water Wholesaler. In *Proc. of the AWWA Annual Conference*. Denver, Colo.: AWWA.

Bellamy, W.D., G.R. Finch, and C.N. Haas. 1998. *Integrated Disinfection Design Framework*. Denver, Colo.: Awwa Research Foundation.

Daniel, P.A. 1998. *Balancing Multiple Water Quality Objectives*. Denver, Colo.: Awwa Research Foundation and AWWA.

Deb, A.K., K.A. Momberger, Y.J. Hasit, and F.M. Grablutz. 2000. *Guidance for Management of Distribution System Operation and Maintenance*. Denver, Colo.: Awwa Research Foundation.

Kirmeyer, G.J., G.W. Foust, G.L. Pierson, J.J. Simmler, and M.W. LeChevallier. 1993. *Optimizing Chloramine Treatment*. Denver, Colo.: Awwa Research Foundation.

Kirmeyer, G.J., M. Friedman, K. Martel, and A. Sandvig. 2002. *Guidance Manual for Monitoring Distribution System Water Quality.* Denver, Colo.: Awwa Research Foundation.

Kirmeyer, G.J., L.H. Odell, J. Jacangelo, A. Wilczak, and R. Wolfe. 1995. *Nitrification Occurrence and Control in Chloraminated Water Systems*. Denver, Colo.: Awwa Research Foundation.

Muylwyk, Q. 2001. Blending Chlorinated and Chloraminated Water in Multiple Distribution Systems. In *Proc. of the AWWA Annual Conference*. Denver, Colo.: AWWA.

Muylwyk, Q., A.L. Smith, and J.A. MacDonald. 1999. Implications on Disinfection Regime When Joining Water Systems: A Case Study of Blending Chlorinated and Chloraminated Water. In *Proc. of the AWWA Water Quality Technology Conference*. Denver, Colo.: AWWA.

Negrin, J. 1998. Using Different Disinfectants in the Same Distribution System. In *Proc. of the AWWA Distribution System Symposium*. Denver, Colo.: AWWA.

SDWA. The Safe Drinking Water Act of 1974. Public Law 93-523. December 16, 1974

———. SDWA Amendments of 1977. Public Law 95-190. November 16, 1977.

Symons, J.M., G.E. Speitel Jr., C. Hwang, S.W. Krasner, and S.E. Barrett. 1998. *Factors Affecting DBP Formation During Chloramination*. Denver, Colo.: Awwa Research Foundation.

Uber, J.G., and R.S. Summers. 2003. *Maintaining Distribution System Residuals Through Booster Chlorination*. Denver, Colo.: Awwa Research Foundation.

US Environmental Protection Agency (USEPA). 1996. *SDWA Amendments of 1996*. Public Law 104-182. Available at: www.epa.gov/OGWDW/sdwa/text.html.

———. 1999. *Guidance Manual for Compliance With the Interim Enhanced Surface Water Treatment Rule: Turbidity Provisions.* EPA-815-R-99-010. Washington, D.C.: USEPA.

This page intentionally blank.

Appendix A

Dechlorination

Dechlorination is the practice of partially or totally removing the chlorine residual. There may be a number of reasons for this practice. The chlorine residual may need to be reduced to a specific level prior to entering the distribution system. Highly chlorinated water used for the disinfection of facilities, pipelines, and appurtenances is commonly dechlorinated upon discharge. Low chlorine residual drinking water must be dechlorinated in some areas before it can be drained into storm sewers or directed to waterways. In addition, water discharged during pipeline break emergencies must sometimes be dechlorinated.

Dechlorination practices have not been optimized to the degree of other operations. Local conditions and regulations have driven the need to dechlorinate and encouraged the development of methods that address these situations. Some of the most important factors that need to be considered include

- flow (or volume) to be dechlorinated,
- chlorine residual level (initial and final),
- time (or distance) that dechlorination must be achieved,
- location,
- chemistry of the water to be dechlorinated,
- impact on the receiving waterway or facility, and
- any regulatory requirements.

System operators should evaluate all of the applicable factors and select the dechlorination method that best suits their situations.

The information provided in this appendix reflects the array of practices currently used by water utilities. The practices described here are not the only methods available but are the most common. In some cases, the performance values presented are from empirical field observations and may not predict the performance of a practice used in another situation. Operators should field test any dechlorination procedure to determine the exact conditions for that application.

Regulatory Issues

Following is a brief description of some of the regulations that impact a decision to dechlorinate. Although there were many individual state and provincial regulations in place at the time this manual was published, reproducing them here would not be useful as they would be almost immediately outdated. It is recommended that system operators stay current regarding applicable regulation requirements. The American Water Works Association (AWWA) and AWWA Sections are good sources of information on current federal, state, and local regulations. The practices described in this manual should be useful for operators to satisfy any dechlorination regulatory requirements.

In some areas, regulations may require a reduction or elimination of the chlorine residual in water prior to discharge. The scope of regulations in some locations can extend to some of the secondary impacts of the treated discharge. For example, overdosing of certain dechlorination chemicals may deplete oxygen concentrations in the receiving streams. Other reactions may raise the pH and, thus, increase the concentration of toxic nonionized ammonia in receiving streams.

Although the need to dechlorinate is not always the result of regulations, it is useful to understand these requirements where they exist. The regulations in the United States and Canada are generally directed at protecting aquatic life. Utilities that are considering the need to dechlorinate should evaluate the potential impact on receiving streams even if there are no enforceable regulations that require this practice.

US Dechlorination Regulations

The US Environmental Protection Agency (USEPA) has established water quality criteria in the Clean Water Act (1986) for total residual chlorine concentration for receiving streams based on acute and chronic toxic effects for aquatic life. Under this requirement, the maximum chlorine concentration is 19 µg/L for the acute toxicity criteria (1 hr average chlorine concentration not more than once every 3 years). The chronic toxicity criterion is 11 µg/L (4-day average not more than every 3 years).

Many states have developed chlorine limits of their own. Permits are required in some cases for the discharge of chlorinated water. Several states limit the residual chlorine concentration to 0.1 mg/L (ppm). Other states have more than one limit for chlorinated water releases. Still other states base the requirements on the water quality criteria of the receiving streams.

Canadian Dechlorination Regulations

The Canadian Environmental Quality Guidelines (1987) propose a water quality criterion of 2 µg/L of total residual chlorine for receiving streams. All provincial regulatory agencies have adopted this as the water quality criterion for receiving streams. Concentration limits for waters discharged into streams vary by province. The allowable chlorine concentrations in these discharges vary from 0.002 to 1 mg/L.

DECHLORINATION PRACTICES

There are a number of nonchemical and chemical methods for dechlorination. The selection of a method depends on site-specific factors such as cost, availability of chemicals, containment logistics, availability of specialized equipment, and regulatory approval. Nonchemical techniques include retention in holding ponds, land application, groundwater recharge, discharge through hay bales and other natural obstructions, and discharge into sanitary sewers. These methods have the advantages of simplicity because they avoid the issues connected with the storage, handling, and safety concerns related to dechlorination chemicals.

Chemicals commonly used for dechlorination include sulfur dioxide, sodium bisulfite, sodium metabisulfite, sodium sulfite, and sodium thiosulfate. Chemicals have advantages over nonchemical methods because they usually require less time to affect dechlorination. However, some of the neutralizing chemicals have potential health concerns if not handled properly.

For example, sulfur dioxide is a hazardous gas. Sodium bisulfite and sodium metabisulfite are skin, eye, or respiratory tract irritants. Sulfite-based chemicals can cause water quality concerns by depleting dissolved oxygen in receiving streams. Some dechlorination chemicals produce hydrochloric acid and decrease water pH. When selecting a chemical for dechlorination, it is important to consider the by-products of the reaction and to receive approval from the appropriate regulatory agency.

Nonchemical Methods

Nonchemical dechlorination methods are generally used for discharges during which the amount of water is limited. The reason for this is that most of these methods require some time to affect the chlorine reduction. They are not suitable for continuous flow treatment. The most common nonchemical methods and estimates of dechlorination effectiveness are listed in Table A-1.

Chemical Dechlorination

Dechlorination is performed by using a (chemical) removing agent to react with the excess oxidizing agent. The reaction between any of the forms of chlorine and the reducing agent is relatively rapid and requires only good mixing to accomplish a satisfactory result. The most commonly used reducing agent (particularly in wastewater treatment plants) is sulfur dioxide, which is a compressed gas available in cylinders similar to those for chlorine and ammonia. Other common reducing agents include sodium bisulfite and sodium sulfite solutions, sodium metabisulfite powder, and sodium thiosulfate. A comparison of the chemical dechlorination agents is provided in Table A-2. The chemical reactions for dechlorination and dechloramination are listed in Tables A-3 and A-4. Activated carbon is included in this section because it is a removing agent, although it is not normally considered a reducing chemical. A brief description of the most common dechlorination agents is provided in the following section.

Table A-1 Nonchemical dechlorination methods

Method of Discharge	Dechlorination Effectiveness Notes
Retention in holding tanks	Chlorine (0.5–2.0 mg/L) reduced in several hours to a few days. Chloramines take three to four times longer.
Land application/flow over pavement or gravel	Chlorine (1.2 mg/L) at 300 gpm reduced to 1.0 mg/L in 500 ft. Chloramines (1–2 mg/L) at 300 gpm needed 2,414 ft to reduce the residual below regulatory requirements.
Groundwater recharge/percolation into groundwater basins	Very effective but groundwater must be carefully monitored. Only approved in limited areas.
Through natural obstructions (hay bales)	May be effective but control is difficult. A crude technique that has limited application.
Storm sewers	Chlorine demand may not be high enough to eliminate a residual prior to discharge to the receiving waterway. May be acceptable depending on the circumstances.
Sanitary sewers	The method most used by water utilities. Must have a sewer entry point near the location of discharge. Need authorization from the sewage agency. Backflow prevention must be practiced.

Table A-2 Comparison of dechlorination agents

Agent	Forms Available	Dose at pH 7.0 (mg/mg Cl)*	pH of 1% Solution	Decomposition/ Off-Gassing	Oxygen Scavenger
Sulfur dioxide	Gas	0.99	Not available	Sulfur dioxide	Strong
Sodium bisulfite	Powder/crystal	1.61	4.3	Sulfur dioxide	Strong
Sodium metabisulfite	Powder/crystal	1.47	4.3	Sulfur dioxide	Strong
Sodium sulfite	Powder/crystal/tablet	1.96	8.5–10.0	Sulfur dioxide	Moderate
Sodium thiosulfate	Powder/crystal	1.9	7.0	Sulfur dioxide	Weak
Calcium thiosulfate	Powder/crystal	1.22	6.5–7.0	Sulfur dioxide	Weak
Ascorbic acid	Powder/crystal/tablet	2.5	6.0*	None	None reported
Sodium ascorbate	Crystal/tablet	2.8	7.0	None	None reported
Hydrogen peroxide	Liquid	0.48	Not available	Oxygen	Adds oxygen

*All dechlorination dosage estimates listed are based on field results (values may differ based on variable conditions). Dechlorination capacity must be checked for each application using site-specific conditions.

Dechlorination using activated carbon. Activated carbon has been widely used in water and wastewater treatment facilities and industries for dechlorination (White 1999). Studies indicate that activated carbon can remove free as well as combined chlorine from water. Granular activated carbon (GAC) is often used for dechlorination activities. In water treatment plants, carbon filters effectively remove dissolved organic matter in addition to removing chlorine. If a significant amount of chlorine reacts with carbon, some of the oxygen attached with carbon may emit as CO or CO_2 gas (White 1999). A fraction of carbon is permanently destroyed during this reaction.

Although carbon can effectively remove chlorine from potable water, several studies indicate that it is more expensive than other dechlorination methods (White 1999; Metcalf and Eddy 1981). In addition, application of carbon for dechlorination activities is not studied extensively.

Sulfur dioxide. Sulfur dioxide (SO_2) is a colorless gas with a suffocating pungent odor. It is widely used in water and wastewater treatment plants for dechlorinating backwash water and wastewater disinfected with chlorine (White 1999).

Dechlorination using sulfur dioxide produces a small amount of acid. Approximately 2.8 mg of alkalinity as calcium carbonate ($CaCO_3$) is consumed per milligram of chlorine reduced. Sulfur dioxide is also an oxygen scavenger. It can deplete dissolved oxygen in the discharge water and in the receiving stream.

Sulfur dioxide is a toxic chemical and it is an extremely irritating gas. It may cause various degrees of irritation to mucous membranes of the eyes, nose, throat, and lungs. Contact with sulfur dioxide liquid may produce freezing of the skin because the liquid absorbs the heat of vaporization from the skin. Concentrations above 500 mg/L can cause acute irritation to the upper respiratory system and cause a sense of suffocation. Great caution is required in transporting cylinders of sulfur dioxide gas.

Care must be taken to design storage and handling facilities to avoid accidental exposure to gas release. While it is suitable for use in facilities such as treatment plants and pumping stations, it is not best suited for field applications. Feed and control systems for sulfur dioxide gas are similar to those for chlorine gas.

Sodium bisulfite. Sodium bisulfite ($NaHSO_3$) is available as a white powder, granule, or clear liquid solution. It is highly soluble in water (39 percent). A 1 percent solution of sodium sulfite has a pH of 4.3 (Southern Ionics 1998). The production of hydrochloric acid (HCl) during chlorine neutralization may marginally decrease pH. Sodium bisulfite is a good oxygen scavenger. Accidental release of slug loads has been reported to have caused injury and death to aquatic species.

Table A-3 Dechlorination chemical reactions with free chlorine

(1a)	C^* + active carbon	$HOCl \rightarrow$ hypochlorous acid	CO^* + surface oxide on carbon	HCl hydrochloric acid	
(1b)	C + carbon	$2Cl_2 + 2H_2O \rightarrow$ chlorine		$4HCl$ + hydrochloric acid	CO_2 carbon dioxide
(2)	$SO_2 + H_2O$ + sulfur dioxide	$HOCl \rightarrow$ hypochlorous acid	H_2SO_4 + sulfuric acid	HCl hydrochloric acid	
(3)	$NaHSO_3$ + sodium bisulfite	$HOCl \rightarrow$ hypochlorous acid	$NaHSO_4$ + sodium bisulfate	HCl hydrochloric acid	
(4)	Na_2SO_3 + sodium sulfite	$HOCl \rightarrow$ hypochlorous acid	Na_2SO_4 + sodium sulfate	HCl hydrochloric acid	
(5)	$Na_2S_2O_5$ + sodium metabisulfite	$2HOCl + H_2O \rightarrow$ hypochlorous acid	$2NaHSO_4$ + sodium bisulfate	$2HCl$ hydrochloric acid	
(6a)	$Na_2S_2O_3$ + sodium thiosulfate	$4HOCl + H_2O \rightarrow$ hypochlorous acid	$2NaHSO_4$ + sodium bisulfate	$4HCl$ hydrochloric acid	
(6b)	$Na_2S_2O_3$ + sodium thiosulfate	$HOCl \rightarrow$ hypochlorous acid	$Na_2SO_4 + S$ + sodium sulfate	HCl hydrochloric acid	
(6c)	$2Na_2S_2O_3$ + sodium thiosulfate	$HOCl \rightarrow$ hypochlorous acid	$Na_2S_4O_6$ + sodium tetrathionate	$NaCl$ + sodium chloride	$NaOH$ sodium hydroxide
(7)	$C_5H_5O_5CH_2OH$ + ascorbic acid	$HOCl \rightarrow$ hypochlorous acid	$C_5H_3O_5CH_2OH$ + dehydroascorbic acid	HCl + hydrochloric acid	H_2O water
(8)	$C_5H_5O_5CH_2ONa$ + sodium ascorbate	$HOCl \rightarrow$ hypochlorous acid	$C_5H_3O_5CH_2OH$ + dehydroascorbic acid	$NaCl$ + hydrochloric acid	H_2O water
(9a)	CaS_2O_3 + calcium thiosulfate	$4HOCl + H_2O \rightarrow$ hypochlorous acid	$CaSO_4$ + calcium sulfate	HCl + hydrochloric acid	H_2SO_4 sulfuric acid
(9b)	CaS_2O_3 + calcium thiosulfate	$HOCl + H_2O \rightarrow$ hypochlorous acid	$CaSO_4$ + calcium sulfate	HCl + hydrochloric acid	S sulfur

Sodium bisulfite is not carcinogenic or mutagenic and is used in food and drugs as a preservative. The US Food and Drug Administration (FDA) recognizes sodium bisulfite as safe when used in accordance with good manufacturing practices or feeding practices. However, sodium bisulfite can cause skin, eye, and respiratory tract irritation. It is harmful if swallowed or inhaled.

The strength of sodium bisulfite solutions diminishes somewhat with age. Sodium bisulfite gradually decomposes in air, producing sulfur dioxide. It reacts strongly with acids to produce sulfur dioxide. Dilution with water also produces sulfur dioxide.

Sodium bisulfite is available only in crystalline/liquid form. This is less convenient than a tablet form for storage, transportation, and handling. However, better control of dosage rates can be obtained by using solutions. Currently, many industries and wastewater utilities use sodium bisulfite solution for dechlorination.

Sodium bisulfite may crystallize at room temperatures. It is highly viscous and sometimes difficult to handle. In addition, sodium bisulfite is highly corrosive and caution must be exercised in safely handling this chemical.

Sodium sulfite. Sodium sulfite (Na_2SO_3) is yet another dechlorinating agent widely used by utilities. It is generally available in powder/crystalline form. In addition, some companies produce sodium sulfite tablets.

Sodium sulfite solutions are slightly alkaline. Fifty grams of sodium sulfite (anhydrous) in 1 L of distilled water produces a solution with a pH of 8.5–10.5. Although it produces HCl, field studies indicated that the reaction does not appreciably decrease solution pH during dechlorination.

Table A-4 Dechlorination chemical reactions with chloramines

(1)	C^* + carbon	$2NH_2Cl + H_2O \rightarrow$ monochloramine	NH_3 + ammonia	HCl + hydrochloric acid	CO^* surface oxide on carbon
(2)	C^* + carbon	$2NHCl_2 + H_2O \rightarrow$ dichloramine	N_2 + nitrogen	4HCl + hydrochloric acid	CO^* surface oxide on carbon
(3)	SO_2 + sulfur dioxide	$NH_2Cl + H_2O \rightarrow$ monochloramine	H_2SO_4 + sulfuric acid	NH_3 + ammonia	HCl hydrochloric acid
(4a)	$Na_2S_2O_3$ + sodium thiosulfate	$4NH_2Cl + 5H_2O \rightarrow$ monochloramine	$2NaHSO_4$ + sodium bisulfate	$4NH_3$ + ammonia	4HCl hydrochloric acid
(4b)	$Na_2S_2O_3$ + sodium thiosulfate	$NH_2Cl + H_2O \rightarrow$ monochloramine	Na_2SO_4 + sodium sulfate	$S + NH_3$ + sulfur ammonia	HCl hypochloric acid
(4c)	$2Na_2S_2O_3$ + sodium thiosulfate	$NH_2Cl + H_2O \rightarrow$ monochloramine	$Na_2S_4O_6$ + sodium tetrathionate	$2NaOH + NH_3$ + sodium hydroxide ammonia	HCl hypochloric acid
(5a)	$Na_2S_2O_3$ + sodium thiosulfate	$2NHCl_2 + 5H_2O \rightarrow$ dichloramine	$2NaHSO_4$ + sodium bisulfate	$2NH_3$ + ammonia	4HCl hypochloric acid
(5b)	$2Na_2S_2O_3$ + sodium thiosulfate	$NHCl_2 + 2H_2O \rightarrow$ dichloramine	$2Na_2SO_4$ + sodium sulfate	$2S + NH_3$ + sulfur ammonia	2HCl hydrochloric acid
(5c)	$4Na_2S_2O_3$ + sodium thiosulfate	$NHCl_2 + 4H_2O \rightarrow$ dichloramine	$2Na_2S_4O_6$ + sodium tetrathionate	$4NaOH + NH_3$ + sodium hydroxide ammonia	2HCl hydrochloric acid
(6)	Na_2SO_3 + sodium sulfite	$NH_2Cl + H_2O \rightarrow$ monochloramine	Na_2SO_4 + sodium sulfate	NH_3 + ammonia	HCl hydrochloric acid
(7)	$2Na_2SO_3$ + sodium sulfite	$NHCl_2 + 2H_2O \rightarrow$ dichloramine	$2Na_2SO_4$ + sodium sulfate	NH_3 + ammonia	2HCl hydrochloric acid
(8)	$NaHSO_3$ + sodium bisulfite	$NH_2Cl + H_2O \rightarrow$ monochloramine	$NaHSO_4$ + sodium bisulfate	NH_3 + ammonia	HCl hydrochloric acid
(9)	$2NaHSO_3$ + sodium bisulfite	$NHCl_2 + 2H_2O \rightarrow$ dichloramine	$2NaHSO_4$ + sodium bisulfate	NH_3 + ammonia	2HCl hydrochloric acid
(10)	$Na_2S_2O_5$ + sodium metabisulfite	$2NH_2Cl + 3H_2O \rightarrow$ monochloramine	$2NaHSO_4$ + sodium bisulfate	$2NH_3$ + ammonia	2HCl hydrochloric acid
(11)	$Na_2S_2O_5$ + sodium metabisulfite	$NHCl_2 + 3H_2O \rightarrow$ dichloramine	$2NaHSO_4$ + sodium bisulfate	NH_3 + ammonia	2HCl hydrochloric acid
(12)	$2C_5H_5O_5CH_2OH$ + ascorbic acid	$2NH_2Cl \rightarrow$ monochloramine	$2C_5H_3O_5CH_2OH$ + dehydroascorbic acid	$2NH_3$ + ammonia	2HCl hydrochloric acid
(13)	$2C_5H_5O_5CH_2ONa$ + sodium ascorbate	$2NH_2Cl \rightarrow$ monochloramine	$2C_5H_3O_5CH_2OH$ + dehydroascorbic acid	$2NH_3$ + ammonia	2NaCl sodium chloride

Sodium sulfite is a reducing agent and is reported to scavenge more oxygen than sodium thiosulfate. Approximately eight parts of sodium sulfite are required to neutralize one part of oxygen. However, field studies indicate that sodium sulfite tablets remove less than 10 percent of dissolved oxygen from waters containing approximately 1 mg/L of residual chlorine.

Sodium sulfite may affect the brain, respiratory system, and skin. It is an eye, skin, mucous membrane, and respiratory tract irritant. Material safety data sheets (MSDSs) indicate crystalline sodium sulfite to be stable. However, sodium sulfite solution will decompose upon reaction with air to form sulfur dioxide gas. The shelf life of sodium sulfite tablets is reported to be approximately 1 year for unopened pails, if stored properly. If opened, the shelf life for the tablets is approximately 2 months.

The major advantage of using sodium sulfite is that currently it is available in tablet form. Many utilities find dechlorination tablets easier to store, handle, and apply as compared to dechlorination solutions or powders.

Sodium metabisulfite.* Sodium metabisulfite ($Na_2S_2O_5$) is available as crystal, powder, or solution. The pH of a 1 percent solution of sodium metabisulfite is 4.3. Production of HCl during neutralization marginally decreases treated water pH. It is a good oxygen scavenger. Its scavenging properties are comparable to that of sodium bisulfite.

Sodium metabisulfite is an eye, throat, skin, and lung irritant. Overexposure to sodium metabisulfite may produce highly toxic effects. Sodium metabisulfite is poison if it enters the body through intravenous route. MSDS warns of adverse reproductive effects due to overexposure. Ingestion may cause mild to moderately severe oral and esophageal burns. Sodium metabisulfite in food can provoke life-threatening asthma.

The stability of sodium metabisulfite increases with concentration. It is slowly degraded when exposed to oxygen. Solutions of 2, 10, and 20 percent strengths are stable for 1, 3, and 4 weeks, respectively. Sodium metabisulfite reacts strongly with acids to produce sulfur dioxide. Dilution with water also produces sulfur dioxide. Sodium metabisulfite decomposes at 150°C and produces sulfur dioxide.

Sodium metabisulfite is available only in crystalline/liquid form. This makes it less convenient for storage, transportation, and handling than a tablet-form dechlorinating agent. Its oxygen scavenging properties are a potential concern in field application.

Sodium thiosulfate. Sodium thiosulfate ($Na_2S_2O_3$) is a colorless crystal. It undergoes various reactions with free chlorine, depending on solution pH (White 1999; General Chemical 1988). The pH of sodium thiosulfate solution is near neutral. Although it produces HCl during dechlorination, field studies have indicated that the reaction does not alter solution pH appreciably. Thiosulfate is an oxygen scavenger and reducing agent. However, it scavenges less oxygen than sodium sulfite, bisulfite, or metabisulfite.

Sodium thiosulfate is a skin, eye, nose, and throat irritant. It is moderately toxic by an intravenous route. A USEPA toxicity study indicated that sodium thiosulfate is not very toxic to aquatic species (Tikkanen et al. 2001).

No published data are available on the stability of sodium thiosulfate. However, utilities using thiosulfate for dechlorination have reported that the strength of thiosulfate solutions does not decrease appreciably after 2 or 3 days of storage.

Sodium thiosulfate reacts with acid to produce sulfur dioxide and hydrogen sulfide (H_2S). It reacts violently with sodium nitrite ($NaNO_2$). When heated to decomposition, it emits toxic fumes of sulfur dioxide and sodium oxide (Na_2O). It reacts rapidly with iron and is readily hydrolyzed by water (i.e., it is hygroscopic).

Sodium thiosulfate is not currently available in tablet form. However, many utilities prefer to use a sodium thiosulfate solution because it is a weaker oxygen scavenger than other dechlorinating agents such as sodium sulfite, bisulfite, and metabisulfite. In addition, the solution allows for better dose control during dechlorination.

A concern over using sodium thiosulfate is that this chemical reacts slowly with chlorine and requires more time for dechlorination than sulfur dioxide and other dechlorination chemicals (White 1999). In addition, over-dechlorination with sodium thiosulfate may encourage thiobacillus and some other bacterial growth in receiving streams, particularly during low flow conditions. A drop in pH, caused by the production of sulfuric acid (H_2SO_4) by microorganisms, has been reported under such conditions.

* $Na_2S_2O_5 + H_2O \rightarrow 2NaHSO_3$ when dissolved.

Other Dechlorination Chemicals

The following chemicals are not widely used for drinking water dechlorination. Information on the use of these chemicals may be limited and local regulatory agency approval may be required prior to considering their use.

Calcium thiosulfate. Calcium thiosulfate (CaS_2O_3) is a clear crystalline substance, with little color, a faintly sulfurous odor, and near neutral pH. Calcium thiosulfate undergoes the following reactions with chlorine (Hardison and Hamamoto 1998).

A 30 percent solution of calcium thiosulfate has a pH of approximately 6.5–7.5. Chlorine neutralization produces HCl and H_2SO_4, which may result in a lower pH. It does not scavenge oxygen and does not produce sulfur dioxide.

Calcium thiosulfate is stable under normal conditions and does not readily release sulfur dioxide gas. However, reaction with acids produces sulfur dioxide. Hazardous reporting is not required. Thermal decomposition may produce sulfur dioxide. Calcium thiosulfate may cause eye and skin irritation.

Calcium thiosulfate does not off-gas sulfur dioxide as other sodium-based dechlorinating agents. It is less toxic to aquatic species. However, dechlorination reactions using stoichiometric concentrations of calcium thiosulfate require nearly 5 min for complete neutralization when it is added (Hardison 2000). Over-dosing of calcium thiosulfate may produce milky-colored suspended solids, causing turbidity violations. Also, excess thiosulfate release may promote thiobacillus bacterial growth. However, bacterial growth is promoted mostly in continuous, excess discharge situations (e.g., cooling water and disinfected wastewater dechlorination operations).

Ascorbic acid (vitamin C). Vitamin C (ascorbic acid) is reported to react with chlorine to produce chloride and dehydroascorbate (Peterka 1998). Because ascorbic acid is weakly acidic, the pH of water may decrease slightly. As reported by Peterka (1998), field application of vitamin C reduces far more chlorine than an equal weight of sulfur-based compound.

The report by Peterka also provides a brief summary on the use of vitamin C for different dechlorination applications. To dechlorinate water from fire hydrants, the use of venturi injectors or other devices, such as the device described in the August 1998 issue of *Opflow,* are recommended. Peterka recommends that care be taken to prevent a drop in pH level below 6.5, particularly during the spawning season or when fingerlings are present. Monitoring of the dissolved oxygen concentration and other water quality parameters are highly recommended.

Ascorbic acid is reasonably stable in a dry state with a shelf life of approximately 3 years (Peterka 1998). However, it rapidly oxidizes in solution. Stability of the solution is affected by concentration and exposure to light and air. A 5 percent weight/volume solution may remain at approximately 95 percent potency level after 12 days, if kept in the dark, whereas a 1 percent solution may remain at approximately 80 percent potency after 10 days. A 0.02 percent solution will degrade to 0 percent within 3 days.

The use of ascorbic acid as a dechlorination chemical is relatively new. There are conflicting reports regarding the amount needed for dechlorination and its effect on pH (Tikkanen et al. 2001). When considering the use of this chemical, utilities should take care to verify these relationships under site-specific conditions. Approval may also be required by regulatory agencies prior to its use.

Sodium ascorbate. Sodium ascorbate is available in crystalline form in 2.5-, 25-, or 100-kg boxes. The chemical is very stable with a shelf life of at least 1 year in a dry state if kept in a cool, dark place. Because sodium ascorbate is more expensive than ascorbic acid and more sodium ascorbate than ascorbic acid is required to neutralize the same amount of chlorine, treatment using sodium ascorbate is significantly more expensive than treatment using ascorbic acid. However, the use of sodium ascorbate may be preferred, as this chemical does not appreciably alter the pH of the discharge or receiving waters. The pH of sodium ascorbate is approximately 7.0.

Utilities considering the use of sodium ascorbate for dechlorination are cautioned that this practice has not been fully field tested. Regulatory agency approval may be required.

Hydrogen peroxide. Hydrogen peroxide (H_2O_2) is yet another chemical that can potentially neutralize chlorine in solution. It reacts with free available chlorine in solutions with a pH greater than 7 according to the following reaction. In most cases, oxygen produced by peroxide will remain dissolved in solution. However, while neutralizing super-chlorinated water in closed environments, the solution may effervesce, and provisions must be made to accommodate the oxygen evolved.

The cost is generally higher than that of sulfur dioxide and other sulfite-based chemicals. Reaction with hydrogen peroxide produces oxygen, which may be beneficial to receiving waters. One of the concerns with the use of hydrogen peroxide is that it is very reactive and rated hazardous when the strength is greater than 52 percent. Hydrogen peroxide at concentrations of 20 percent or greater is rated as an "oxidizer and corrosive" by the US Department of Transportation (USDOT) and must be labeled accordingly.

Due to USDOT and Superfund Amendments and Reauthorization Act (SARA) Title III regulatory requirements, hydrogen peroxide may not be the best dechlorination choice for field applications. It may, however, be a viable alternative to sulfur dioxide for industrial, water, and wastewater treatment plant applications. The use of hydrogen peroxide as a dechlorination chemical has not been fully field tested. Regulatory approval may be required prior to use.

CHEMICAL FEED TECHNIQUES

There are several methods commonly used for feeding dechlorination chemicals. Sulfur dioxide gas is the preferred method for use in treatment plants and pump stations, the other methods are used both in facilities and in the field.

Sulfur Dioxide Gas Feed Facilities

The components of a sulfur dioxide gas feed system are identical to those described for chlorine gas feed systems (chapter 4) (White 1999). However, there are some differences in the procedures for feeding and handling this chemical.

Sulfur dioxide has a very low vapor pressure. This can cause problems with reliquefaction. Gas cylinders and storage tanks may be heated to help avoid this occurrence. Pressure control valves are also employed upon withdrawal to eliminate reliquefaction. Liquid withdrawal systems include evaporators and pressure-reducing valves as well.

Piping and other system components can be the same as for chlorine systems. One exception is that line and auxiliary valves can be 316 SS, rather than monel. Accessory equipment, safety equipment, automatic cylinder switchover equipment, remote vacuum equipment, flexible connections, sulfonators, evaporators, and housing requirements are all the same as for chlorine gas systems with only minor modifications.

Sulfur dioxide and chlorine gas systems are always separated. Piping is kept separated and areas for the supply systems and storage rooms are kept apart.

Injectors for sulfur dioxide are usually those used for chlorine although this causes them to be oversized because sulfur dioxide is much more soluble in water than chlorine. Diffusers are usually 316 SS. Mixing and contact time are not as critical for sulfur dioxide due to the difference in solubility.

Sulfur dioxide gas detectors are needed. These are very sensitive and should be connected to a proper alarm system. Sulfur dioxide is a toxic gas. Chlorine residual monitors should be employed to ensure the desired dechlorination result is obtained.

Solid and Liquid Chemical Feed Methods

Gravity feed method. The gravity feed method typically involves adding dechlorinating solution from a container equipped with a spigot that would be placed on the curb above the water flow path. The discharge spigot on the container can be adjusted to provide a minimum dechlorinating solution feed rate into the water flow, based on calculations involving the concentration of the dechlorinating solution, water flow rate, and residual chlorine concentration in the flow stream.

In order to minimize the volumes of dechlorinating solution needed on field vehicles, dry dechlorinating agent should be mixed directly within the container prior to use wherever possible, rather than using premixed dechlorinating solutions provided by suppliers. As part of the dechlorination procedure, samples may be collected downstream of the feed point and analyzed for pH, dissolved oxygen, and residual chlorine, and, if necessary, the chemical feed rate can be adjusted to ensure a nondetectable residual chlorine concentration prior to discharge to the surface water.

Gravity feed systems are simple to operate, have minimal equipment requirements, have been used effectively by various utilities, and are inexpensive (low-density polyethylene carboys equipped with spigots and having a capacity of 5.5 gal). Unless adjusted, chemical feed rates can be expected to decrease slightly over time, as the available head pressure within the carboy decreases during use. A disadvantage of using this technique is that it involves field testing and calculations for flow rates and water quality parameters to adjust chemical feed rate. Field maintenance crews sometimes prefer a method that does not involve field calculations.

Chemical metering pumps. This method for injecting a dechlorinating solution is similar to the gravity feed method, except that a chemical metering pump is used to inject the dechlorinating solution from a container into the water flow. Chemical metering pumps are capable of delivering chemical solutions over a wide range of flows (e.g., 0.0006–3,400 mL/min), the flow rates are adjustable, and the pumps provide a constant chemical feed rate.

Relative to a gravity feed system, this type of feed system is more difficult to operate, requires more equipment (e.g., storage container, pump, energy source, and tubing), and costs are significantly higher. Although chemical metering pumps may provide field personnel with a more reproducible method, this feature involves much higher cost and greater operator ability and attention during dechlorination.

Venturi injector systems. Venturi injectors (Figure A-1) are differential pressure injection devices that allow for the injection of liquids (e.g., dechlorinating solutions from hydrants) into a pressurized water stream. This method is well suited for pressurized water releases such as those through hydrants. Pressurized water entering the injector inlet is constricted toward the injection chamber. This results in a higher velocity water stream through the injection chamber than at the injector inlet. The increase in velocity through the injection chamber results in a decrease in pressure, thereby allowing the dechlorinating agent to be drawn from a storage container through the suction port and entrained into the water stream.

Water main discharges would be constricted by a regulator valve or gate valve and routed through a venturi injector system. Dechlorinating solution would be drawn into the injector from a plastic container with the chemical feed rate controlled by a metering valve. In addition to the valves, venturi injector units should include a flowmeter installed near the metering valve to measure the chemical feed rate (e.g., a rotameter), a threaded fitting at the upstream end of the pipe to attach adapters (e.g., reducing sections), and a fitting at the downstream end of the pipe for attachment of flexible discharge piping (e.g., a hose).

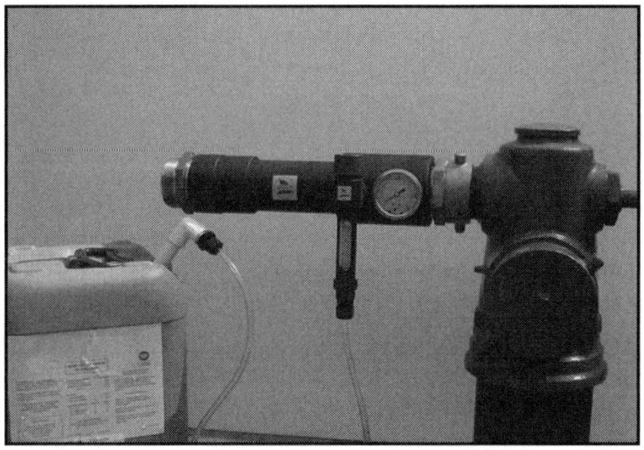

Courtesy of Arden Industries.

Figure A-1 "Bazooka" venturi dechlorination feeder

The primary advantages of venturi injection systems include

- efficient operation over a wide range of pressures;
- available in a wide range of sizes, flows, and injection capacities;
- no external energy requirements; and
- instantaneous mixing via cavitation in the injection chamber.

The primary disadvantages of venturi injection systems include

- more sophisticated equipment requirements than gravity feed systems,
- slightly more labor-intensive set-up,
- higher unit equipment cost than gravity feed systems, and
- may require constant monitoring.

Spray feed systems. The dechlorinating solution may be sprayed into the flow (on pipe walls, surfaces, or pipeline appurtenances) via a backpack sprayer (similar to those used for pesticide and herbicide application) or chemical fire extinguisher. The advantages of this technique are that the chemical feed rate is fairly constant (given a steady pressure within the solution chamber). Hence, dosages can be approximated fairly accurately, and a piped or channelized flow is not required to effectively feed the chemical. This method is typically more effective in adding dechlorinating chemicals to sheet flows than the other alternatives evaluated earlier.

A significant disadvantage of a spray feed system is that it requires the equipment to be set up at a stationary point and monitored for adequate chemical and pressure. If a stationary point is not available, a person will have to don the sprayer and continuously apply the dechlorinating agent to the flow, which is very labor intensive.

Flow-through systems. Flow-through systems (Figures A-2 and A-3) include any method where the solid chemical is held stationary and the flow allowed to run over, around, and/or through it. Examples include pumping chlorinated water through a container filled with dechlorinating agent or laying permeable bags of the chemical in the flow path. For this application, it is anticipated that the dechlorinating agent would be used in tablet or powder form. The advantages of flow-through systems are that they are simple and can be used for sheet flow applications as well as for channelized or pumped flow. The disadvantages are that there is no control over the dosage. Overdosing or underdosing to significant levels could easily occur and it may

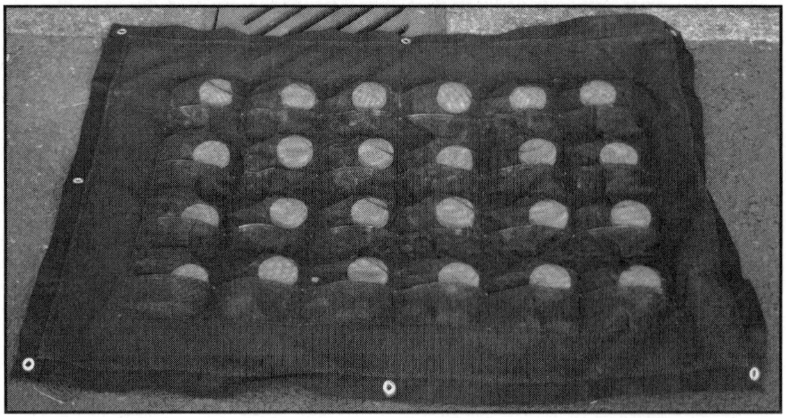

Source: Pollard Water dot com.

Figure A-2 Mat for dechlorination of trenches during main breaks

Source: DAVCO Associates.

Figure A-3 Diffuser for dechlorination of hydrant or blowoff waters

be difficult, in some cases, to tell when the chemical has been used up and must be replaced. Also, due to the contact time required for dissolution of powder/tablet, this method is more suitable for low- and medium-velocity discharges. This method may not be suitable for dechlorinating releases from unidirectional flushing where the velocity of the flow is approximately 5.0 ft/s (1.5 m/s). Some variations in the application of flow-through systems are described in the following section.

Automatic tablet dispensers. Automatic tablet feeders for dechlorination are similar to the tablet feeders currently used for disinfection in many water/wastewater treatment plants. The feeders typically consist of a nonmoving housing and automatic feed tubes. The tubes are inserted down through a removable top cover of the feeder into the stream of water. The lower end of each tube is slotted to permit free flow of water through the tubes to ensure good contact between the water and dechlorination tablets. The feeders typically contain a removable wire plate at the outlet end to control the internal water level based on the flow rate and level of dechlorination required. Proper connection is required to direct the flow to the feeders, which typically have a 6-in. pipe inlet or solid inlet end for field adaptation.

As the stream of water flows past the feed tubes containing dechlorination tablets, the dechlorination agent is released into the water by dissolution. At the outlet end, a weir controls the height of the water level in the feeder, which controls the concentration of the agent in the water, regardless of surges in the water flow entering the feeder. As the incoming water flow rate increases, the water level in the unit rises, immersing a greater number of tablets. Because the amount of agent dissolved depends upon the number of tablets immersed in the water, the dechlorination agent level remains constant regardless of the water level in the feeder.

The feeders commonly used by utilities and industrial facilities that provided information for this report are built with medium-density polyethylene, with the following dimensions: 26 in. long, 18 in. wide, and 16 in. deep. Each feeder is equipped with up to four feeder tubes, 24 in. long with an outside diameter of 3.5 in. The feeders can handle up to an average flow rate of 35 gpm (0.002 m^3/s) and a peak flow (4 hr) of 87 gpm (0.005 m^3/s). Hence, they are better suited for dechlorination of smaller chlorinated water discharges. Multiple feeders can be used for dechlorinating higher flows.

Flow Control Measures

During planned and unplanned water releases, it may be necessary to construct flow control measures to prevent the water from entering directly into a water body and to provide an opportunity for better mixing of the dechlorinating agent. Construction of berms, swales, ditches, or redirection pipes to control the flow of released water is recommended.

Berms. Berms can be constructed using sandbags, hay bales, gravel with a filter fabric core, plywood, or similar materials. Sandbags are often used by utilities to construct temporary berms. Sandbags placed in a semicircle, with a depression in the ground upstream, may provide a pond-like structure with a higher residence time for dechlorination. Sandbags can be made from material available at site or brought to site.

Swale. A swale, or natural depression near the point of chlorinated water release, can be used to control the flow of water. If such a setting is available near the flow, the water can be redirected toward it. Constructing a berm at one end of the swale could significantly increase the holding capacity.

Ditches. Directing chlorinated water flow to ditches nearby to allow for better mixing and dechlorination may be effective. If available, ditches near chlorinated water release locations can be used as temporary holding ponds. Check dams can be constructed using sand bags or hay bales to improve mixing. The flow of water can be directed toward the ditches using sandbags and hay bales.

However, caution must be exercised in using this technique because, in many places, ditches may be considered to be waters of the state for which water quality standards apply. Clearly, this approach would not be acceptable in such a situation.

Redirection pipe. A redirection pipe may be used to redirect the flow of water to a specific area or into a holding structure. It could also be used as a treatment system if an in-line injection system is used to inject the dechlorinating agent into the pipe.

Polyvinyl chloride (PVC) pipe is recommended for constructing redirection pipes. These pipes are not expensive and are readily available in different diameters. If the in-line injection method is used, there should be an injection point at the end closest to the source of the chlorinated water. This permits the dechlorinating agent to mix within the redirection pipe prior to entering the receiving water.

The injection point can be a vertical pipe that intercepts the discharge pipe. It can be constructed by fitting a T-joint into the pipe. The vertical pipe should be short enough (less than 1 m) to minimize lifting of chemicals by the operators during the process. A chemical metering pump can be fitted to the top of the injection point to control chemical dose rate.

REFERENCES

General Chemical. 1988. *Product Bulletin: Water Treatment Dechlorination*. Parsippany, N.J.: General Chemical.

Hardison, R.L., 2000. *A New and Better Dechlorination Process for Wastewater and Drinking Water.* Proc. Small Drinking Water and Wastewater Systems International Symposium and Technology Expo, Phoenix., Ariz.

Hardison, R.L., and M. Hamamoto. 1998. *An Alternative Dechlorination Process*. Oakland, Calif.: CWEA (California Water Environment Association) Annual Conference.

Metcalf, A., and M.S. Eddy. 1981. *Wastewater Engineering: Treatment, Disposal, and Reuse*. New York: McGraw Hill.

Peterka, G. 1998. Vitamin C: A Promising Dechlorination Reagent. *Opflow,* 24(12):1–5.

SARA (Superfund Amendments and Reauthorization Act, Title III). 1986. Emergency Planning and Community Right to Know Act. 42 U.S.C. 11001 et seq.

Southern Ionics, Inc. 1998. *Application Bulletin: Sodium Bisulfite Solution*. Matthews, N.C.: Southern Ionics, Inc.

Tikkanen M., J. Schroeter, L. Leong, and R. Ganesh. 2001. *Guidance Manual for Disposal of Chlorinated Water.* Denver, Colo.: Awwa Research Foundation.

White, G.C. 1999. *Handbook of Chlorination and Alternative Disinfectants*. New York: John Wiley & Sons.

Appendix B

CT Values for Inactivation of *Giardia* and Viruses by Free Chlorine and Other Disinfectants

All tables in this appendix are taken from *Guidance Manual for Compliance with the Filtration and Disinfection Requirements for Public Water Systems Using Surface Water Sources*, Appendix E, Science and Technology Branch, Criteria and Standards Division, Office of Drinking Water, USEPA, Washington, D.C., October 1989.

CT Values for Inactivation of *Giardia* Cysts by Free Chlorine at 0.5°C or Lower

CHLORINE CONCENTRATION	pH<=6 Log Inactivation						pH=6.5 Log Inactivation						pH=7.0 Log Inactivation						pH=7.5 Log Inactivation					
(mg/L)	0.5	1.0	1.5	2.0	2.5	3.0	0.5	1.0	1.5	2.0	2.5	3.0	0.5	1.0	1.5	2.0	2.5	3.0	0.5	1.0	1.5	2.0	2.5	3.0
<=0.4	23	46	69	91	114	137	27	54	82	109	136	163	33	65	98	130	163	195	40	79	119	158	198	237
0.6	24	47	71	94	118	141	28	56	84	112	140	169	33	67	100	133	167	200	40	80	120	159	199	239
0.8	24	48	73	97	121	145	29	57	86	115	143	172	34	68	103	137	171	205	41	82	123	164	205	246
1	25	49	74	99	123	148	29	59	88	117	147	176	35	70	105	140	175	210	42	84	127	169	211	253
1.2	25	51	76	101	127	152	30	60	90	120	150	180	36	72	108	143	179	215	43	86	130	173	216	259
1.4	26	52	78	103	129	155	31	61	92	123	153	184	37	74	111	147	184	221	44	89	133	177	222	266
1.6	26	52	79	105	131	157	32	63	95	126	155	189	38	75	113	151	188	226	46	91	137	182	228	273
1.8	27	54	81	108	135	162	32	64	97	129	161	193	39	77	116	154	193	231	47	93	140	186	233	279
2	28	55	83	110	138	165	33	66	99	131	164	197	39	79	118	157	197	236	48	95	143	191	238	286
2.2	28	56	85	113	141	169	34	67	101	134	169	201	40	81	121	161	202	242	50	99	149	198	248	297
2.4	29	57	86	115	143	172	34	68	103	137	171	205	41	82	124	165	206	247	50	99	149	199	248	298
2.6	29	58	88	117	146	175	35	70	105	139	174	209	42	84	126	168	210	252	51	101	152	203	253	304
2.8	30	59	89	119	148	178	36	71	107	142	178	213	43	86	129	171	214	257	52	103	155	207	258	310
3	30	60	91	121	151	181	36	72	109	145	181	217	44	87	131	174	218	261	53	105	158	211	263	316

CHLORINE CONCENTRATION	pH=8.0 Log Inactivation						pH=8.5 Log Inactivation						pH=9.0 Log Inactivation					
(mg/L)	0.5	1.0	1.5	2.0	2.5	3.0	0.5	1.0	1.5	2.0	2.5	3.0	0.5	1.0	1.5	2.0	2.5	3.0
<=0.4	46	92	139	185	231	277	55	110	165	219	274	329	65	130	195	260	325	390
0.6	48	95	143	191	238	286	57	114	171	228	285	342	68	136	204	271	339	407
0.8	49	98	148	197	246	295	59	113	177	236	295	354	70	141	211	281	352	422
1	51	101	152	203	253	304	61	122	183	243	304	365	73	146	219	291	364	437
1.2	52	104	157	209	261	313	63	125	188	251	313	376	75	150	226	301	376	451
1.4	54	107	161	214	268	321	65	129	194	258	323	387	77	155	232	309	387	464
1.6	55	110	165	219	274	329	66	132	199	265	331	397	80	159	239	318	398	477
1.8	56	113	169	225	282	338	68	136	204	271	339	407	82	163	245	326	408	489
2	55	115	173	231	288	346	70	139	209	278	348	417	83	167	250	333	417	500
2.2	59	118	177	235	294	353	71	142	213	284	355	426	85	170	256	341	426	511
2.4	60	120	181	241	301	361	73	145	218	290	363	435	87	174	261	348	435	522
2.6	61	123	184	245	307	368	74	148	222	296	370	444	89	178	267	355	444	533
2.8	63	125	188	250	313	375	75	151	226	301	377	452	91	181	272	362	453	543
3	64	127	191	255	318	382	77	153	230	307	383	460	92	184	276	369	460	552

CT Values for Inactivation of *Giardia* Cysts by Free Chlorine at 5°C

CHLORINE CONCENTRATION	pH<=6 Log Inactivation						pH=6.5 Log Inactivation						pH=7.0 Log Inactivation						pH=7.5 Log Inactivation					
(mg/L)	0.5	1.0	1.5	2.0	2.5	3.0	0.5	1.0	1.5	2.0	2.5	3.0	0.5	1.0	1.5	2.0	2.5	3.0	0.5	1.0	1.5	2.0	2.5	3.0
<=0.4	16	32	49	65	81	97	20	39	59	78	98	117	23	46	70	93	116	139	28	55	83	111	138	166
0.6	17	33	50	67	83	100	20	40	60	80	100	120	24	49	72	95	119	143	29	57	86	114	143	171
0.8	17	34	52	69	86	103	20	41	61	81	102	122	24	49	73	97	122	146	29	58	88	117	146	175
1	18	35	53	70	88	105	21	42	63	83	104	125	25	50	75	99	124	149	30	60	90	119	149	179
1.2	18	36	54	71	89	107	21	42	64	85	106	127	25	51	76	101	127	152	31	61	92	122	153	183
1.4	18	36	55	73	91	109	22	43	65	97	108	130	26	52	78	103	129	155	31	62	94	125	156	187
1.6	19	37	56	74	93	111	22	44	66	88	110	132	26	53	79	105	132	158	32	64	96	128	160	192
1.8	19	38	57	76	95	114	23	45	69	90	113	135	27	54	81	108	135	162	33	65	98	131	163	196
2	19	39	58	77	97	116	23	46	69	92	115	138	28	55	83	110	138	165	33	67	100	133	167	200
2.2	20	39	59	79	98	118	23	47	70	93	117	140	28	56	85	113	141	169	34	68	102	136	170	204
2.4	20	40	60	80	100	120	24	48	72	95	119	143	29	57	86	115	143	172	35	70	105	139	174	209
2.6	20	41	61	81	102	122	24	49	73	97	122	146	29	58	88	117	146	175	36	71	107	142	178	213
2.8	21	41	62	83	103	124	25	49	74	99	123	148	30	59	89	119	148	178	36	72	109	145	181	217
3	21	42	63	84	105	126	25	50	76	101	126	151	30	61	91	121	152	182	37	74	111	147	184	221

CHLORINE CONCENTRATION	pH=8.0 Log Inactivation						pH=8.5 Log Inactivation						pH=9.0 Log Inactivation					
(mg/L)	0.5	1.0	1.5	2.0	2.5	3.0	0.5	1.0	1.5	2.0	2.5	3.0	0.5	1.0	1.5	2.0	2.5	3.0
<=0.4	33	66	99	132	165	198	39	79	118	157	197	236	47	93	140	186	233	279
0.6	34	68	102	136	170	204	41	81	122	163	203	244	49	97	146	194	243	291
0.8	35	70	105	140	175	210	42	84	126	168	210	252	50	100	151	201	251	301
1	36	72	108	144	180	216	43	87	130	173	217	260	52	104	156	208	260	312
1.2	37	74	111	147	184	221	45	89	134	178	223	267	53	107	160	213	267	320
1.4	38	76	114	151	189	227	46	91	137	183	228	274	55	110	165	219	274	329
1.6	39	77	116	155	193	232	47	94	141	197	234	281	56	112	169	225	281	337
1.8	40	79	119	159	198	238	48	96	144	191	239	287	58	115	173	230	288	345
2	41	81	122	162	203	243	49	98	147	196	245	294	59	118	177	235	294	353
2.2	41	83	124	165	207	248	50	100	150	200	250	300	60	120	181	241	301	361
2.4	42	84	127	169	211	253	51	102	153	204	255	306	61	123	184	245	307	368
2.6	43	86	129	172	215	258	52	104	156	208	260	312	63	125	189	250	313	375
2.8	44	88	132	175	219	263	53	106	159	212	265	318	64	127	191	255	318	382
3	45	89	134	179	223	268	54	108	162	216	270	324	65	130	195	259	324	389

CT Values for Inactivation of *Giardia* Cysts by Free Chlorine at 10°C

CHLORINE CONCENTRATION (mg/L)	pH<=6 Log Inactivation						pH=6.5 Log Inactivation						pH=7.0 Log Inactivation						pH=7.5 Log Inactivation					
	0.5	1.0	1.5	2.0	2.5	3.0	0.5	1.0	1.5	2.0	2.5	3.0	0.5	1.0	1.5	2.0	2.5	3.0	0.5	1.0	1.5	2.0	2.5	3.0
<=0.4	12	24	37	49	61	73	15	29	44	59	73	88	17	35	52	69	87	104	21	42	63	83	104	125
0.6	13	25	38	50	63	75	15	30	45	60	75	90	18	36	54	71	89	107	21	43	64	85	107	128
0.8	13	26	39	52	65	78	15	31	46	61	77	92	18	37	55	73	92	110	22	44	66	87	109	131
1	13	26	40	53	66	79	16	31	47	63	78	94	19	37	56	75	93	112	22	45	67	89	112	134
1.2	13	27	40	53	67	80	16	32	48	63	79	95	19	38	57	76	95	114	23	46	69	91	114	137
1.4	14	27	41	55	68	82	16	33	49	65	82	98	19	39	58	77	97	116	23	47	70	93	117	140
1.6	14	28	42	55	69	83	17	33	50	66	83	99	20	40	60	79	99	119	24	48	72	96	120	144
1.8	14	29	43	57	72	86	17	34	51	67	84	101	20	41	61	81	102	122	25	49	74	98	123	147
2	15	29	44	58	73	87	17	35	52	69	87	104	21	41	62	83	103	124	25	50	75	100	125	150
2.2	15	30	45	59	74	89	18	35	53	70	88	105	21	42	64	85	106	127	26	51	77	102	128	153
2.4	15	30	45	60	75	90	18	36	54	71	89	107	22	43	65	86	108	129	26	52	79	105	131	157
2.6	15	31	46	61	77	92	18	37	55	73	92	110	22	44	66	87	109	131	27	53	80	107	133	160
2.8	16	31	47	62	78	93	19	37	56	74	93	111	22	45	67	89	112	134	27	54	82	109	136	163
3	16	32	48	63	79	95	19	38	57	75	94	113	23	46	69	91	114	137	28	55	83	111	138	166

CHLORINE CONCENTRATION (mg/L)	pH=8.0 Log Inactivation						pH=8.5 Log Inactivation						pH=9.0 Log Inactivation					
	0.5	1.0	1.5	2.0	2.5	3.0	0.5	1.0	1.5	2.0	2.5	3.0	0.5	1.0	1.5	2.0	2.5	3.0
<=0.4	25	50	75	99	124	149	30	59	89	118	148	177	35	70	105	139	174	209
0.6	26	51	77	102	128	153	31	61	92	122	153	183	36	73	109	145	182	218
0.8	26	53	79	105	132	158	32	63	95	126	158	189	38	75	113	151	188	226
1	27	54	81	108	135	162	33	65	98	130	163	195	39	78	117	156	195	234
1.2	28	55	83	111	138	166	33	67	100	133	167	200	40	80	120	160	200	240
1.4	28	57	85	113	142	170	34	69	103	137	172	206	41	82	124	165	206	247
1.6	29	58	87	116	145	174	35	70	106	141	176	211	42	84	127	169	211	253
1.8	30	60	90	119	149	179	36	72	108	143	179	215	43	86	130	173	216	259
2	30	61	91	121	152	182	37	74	111	147	184	221	44	88	133	177	221	265
2.2	31	62	93	124	155	186	38	75	113	150	188	225	45	90	136	181	226	271
2.4	32	63	95	127	158	190	38	77	115	153	192	230	46	92	138	184	230	276
2.6	32	65	97	129	162	194	39	78	117	156	195	234	47	94	141	187	234	281
2.8	33	66	99	131	164	197	40	80	120	159	199	239	48	96	144	191	239	287
3	34	67	101	134	168	201	41	81	122	162	203	243	49	97	146	195	243	292

CT Values for Inactivation of *Giardia* Cysts by Free Chlorine at 15°C

CHLORINE CONCENTRATION (mg/L)	pH<=6 Log Inactivation						pH=6.5 Log Inactivation						pH=7.0 Log Inactivation						pH=7.5 Log Inactivation					
	0.5	1.0	1.5	2.0	2.5	3.0	0.5	1.0	1.5	2.0	2.5	3.0	0.5	1.0	1.5	2.0	2.5	3.0	0.5	1.0	1.5	2.0	2.5	3.0
<=0.4	8	16	25	33	41	49	10	20	30	39	49	59	12	23	35	47	58	70	14	28	42	55	69	83
0.6	8	17	25	33	42	50	10	20	30	40	50	60	12	24	36	48	60	72	14	29	43	57	72	86
0.8	9	17	26	35	43	52	10	20	31	41	51	61	12	24	37	49	61	73	15	29	44	59	73	88
1	9	18	27	35	44	53	11	21	32	42	53	63	13	25	38	50	63	75	15	30	45	60	75	90
1.2	9	18	27	36	45	54	11	21	32	43	53	64	13	25	38	51	63	76	15	31	46	61	77	92
1.4	9	18	28	37	46	55	11	22	33	43	54	65	13	26	39	52	65	78	16	31	47	63	78	94
1.6	9	19	28	37	47	56	11	22	33	44	55	66	13	26	40	53	66	79	16	32	48	64	80	96
1.8	10	19	29	38	48	57	11	23	34	45	57	68	14	27	41	54	68	81	16	33	49	65	82	98
2	10	19	29	39	48	58	12	23	35	46	58	69	14	28	42	55	69	83	17	33	50	67	83	100
2.2	10	20	30	39	49	59	12	23	35	47	58	70	14	28	43	57	71	85	17	34	51	68	85	102
2.4	10	20	30	40	50	60	12	24	36	48	60	72	14	29	43	57	72	86	18	35	53	70	88	105
2.6	10	20	31	41	51	61	12	24	37	49	61	73	15	29	44	59	73	88	18	36	54	71	89	107
2.8	10	21	31	41	52	62	12	25	37	49	62	74	15	30	45	59	74	89	18	36	55	73	91	109
3	11	21	32	42	53	63	13	25	38	51	63	76	15	30	46	61	76	91	19	37	56	74	93	111

CHLORINE CONCENTRATION (mg/L)	pH=8.0 Log Inactivation						pH=8.5 Log Inactivation						pH=9.0 Log Inactivation					
	0.5	1.0	1.5	2.0	2.5	3.0	0.5	1.0	1.5	2.0	2.5	3.0	0.5	1.0	1.5	2.0	2.5	3.0
<=0.4	17	33	50	66	83	99	20	39	59	79	98	118	23	47	70	93	117	140
0.6	17	34	51	68	85	102	20	41	61	81	102	122	24	49	73	97	122	146
0.8	18	35	53	70	88	105	21	42	63	84	105	126	25	50	76	101	126	151
1	18	36	54	72	90	108	22	43	65	87	108	130	26	52	78	104	130	156
1.2	19	37	56	74	93	111	22	45	67	89	112	134	27	53	80	107	133	160
1.4	19	38	57	76	95	114	23	46	69	91	114	137	28	55	83	110	138	165
1.6	19	39	58	77	97	116	24	47	71	94	118	141	28	56	85	113	141	169
1.8	20	40	60	79	99	119	24	48	72	96	120	144	29	59	87	115	144	173
2	20	41	61	81	102	122	25	49	74	98	123	147	30	59	89	118	148	177
2.2	21	41	62	83	103	124	25	50	75	100	125	150	30	60	91	121	151	181
2.4	21	42	64	85	106	127	26	51	77	102	128	153	31	61	92	123	153	184
2.6	22	43	65	86	108	129	26	52	78	104	130	156	31	63	94	125	157	188
2.8	22	44	66	88	110	132	27	53	80	106	133	159	32	64	96	127	159	191
3	22	45	67	89	112	134	27	54	81	109	135	162	33	65	98	130	163	195

CT Values for Inactivation of *Giardia* Cysts by Free Chlorine at 20°C

CHLORINE CONCENTRATION (mg/L)	pH<=6 Log Inactivation						pH=6.5 Log Inactivation						pH=7.0 Log Inactivation						pH=7.5 Log Inactivation					
	0.5	1.0	1.5	2.0	2.5	3.0	0.5	1.0	1.5	2.0	2.5	3.0	0.5	1.0	1.5	2.0	2.5	3.0	0.5	1.0	1.5	2.0	2.5	3.0
<=0.4	6	12	18	24	30	36	7	15	22	29	37	44	9	17	26	35	43	52	10	21	31	41	52	62
0.6	6	13	19	25	32	38	8	15	23	30	38	45	9	18	27	36	45	54	11	21	32	43	53	64
0.8	7	13	20	26	33	39	8	15	23	31	38	46	9	18	28	37	46	55	11	22	33	44	55	66
1	7	13	20	26	33	39	8	16	24	31	39	47	9	19	28	37	47	56	11	22	34	45	56	67
1.2	7	13	20	27	33	40	8	16	24	32	40	48	10	19	29	38	48	57	12	23	35	46	58	69
1.4	7	14	21	27	34	41	8	16	25	33	41	49	10	19	29	39	48	58	12	23	35	47	58	70
1.6	7	14	21	28	35	42	8	17	25	33	42	50	10	20	30	39	49	59	12	24	36	48	60	72
1.8	7	14	22	29	36	43	9	17	26	34	43	51	10	20	31	41	51	61	12	25	37	49	62	74
2	7	15	22	29	37	44	9	17	26	35	43	52	10	21	31	41	52	62	13	25	38	50	63	75
2.2	7	15	22	29	37	44	9	18	27	35	44	53	11	21	32	42	53	63	13	26	39	51	64	77
2.4	8	15	23	30	38	45	9	18	27	36	45	54	11	22	33	43	54	65	13	26	39	52	65	78
2.6	8	15	23	31	38	46	9	18	28	37	46	55	11	22	33	44	55	66	13	27	40	53	67	80
2.8	8	16	24	31	39	47	9	19	28	37	47	56	11	22	34	45	56	67	14	27	41	54	68	81
3	9	16	24	31	39	47	10	19	29	38	48	57	11	23	34	45	57	68	14	28	42	55	69	83

CHLORINE CONCENTRATION (mg/L)	pH=8.0 Log Inactivation						pH=8.5 Log Inactivation						pH=9.0 Log Inactivation					
	0.5	1.0	1.5	2.0	2.5	3.0	0.5	1.0	1.5	2.0	2.5	3.0	0.5	1.0	1.5	2.0	2.5	3.0
<=0.4	12	25	37	49	62	74	15	30	45	59	74	89	19	35	53	70	88	105
0.6	13	26	39	51	64	77	15	31	46	61	77	92	18	36	55	73	91	109
0.8	13	26	40	53	66	79	16	32	48	63	79	95	19	38	57	75	94	113
1	14	27	41	54	68	81	16	33	49	65	82	98	20	39	59	78	98	117
1.2	14	28	42	55	69	83	17	33	50	67	83	100	20	40	60	80	100	120
1.4	14	28	43	57	71	85	17	34	52	69	86	103	21	41	62	82	103	123
1.6	15	29	44	58	73	87	18	35	53	70	88	105	21	42	63	84	105	126
1.8	15	30	45	59	74	89	18	36	54	72	90	108	22	43	65	86	108	129
2	15	30	46	61	76	91	18	37	55	73	92	110	22	44	66	88	110	132
2.2	16	31	47	62	78	93	19	38	57	75	94	113	23	45	68	90	113	135
2.4	16	32	48	63	79	95	19	38	58	77	96	115	23	46	69	92	115	139
2.6	16	32	49	65	81	97	20	39	59	78	98	117	24	47	71	94	117	141
2.8	17	33	50	66	83	99	20	40	60	79	99	119	24	48	72	95	119	143
3	17	34	51	67	84	101	20	41	61	81	102	122	24	49	73	97	122	146

CT Values for Inactivation of *Giardia* Cysts by Free Chlorine at 25°C

CHLORINE CONCENTRATION (mg/L)	pH<=6 Log Inactivation						pH=6.5 Log Inactivation						pH=7.0 Log Inactivation						pH=7.5 Log Inactivation					
	0.5	1.0	1.5	2.0	2.5	3.0	0.5	1.0	1.5	2.0	2.5	3.0	0.5	1.0	1.5	2.0	2.5	3.0	0.5	1.0	1.5	2.0	2.5	3.0
<=0.4	4	8	12	16	20	24	5	10	15	19	24	29	6	12	18	23	29	35	7	14	21	28	35	42
0.6	4	8	13	17	21	25	5	10	15	20	25	30	6	12	18	24	30	36	7	14	22	29	36	43
0.8	4	9	13	17	22	26	5	10	16	21	26	31	6	12	19	25	31	37	7	15	22	29	37	44
1	4	9	13	17	22	26	5	10	16	21	26	31	6	12	19	25	31	37	8	15	23	30	38	45
1.2	5	9	14	18	23	27	5	11	16	21	27	32	6	13	19	25	32	38	8	15	23	31	38	46
1.4	5	9	14	18	23	27	6	11	17	22	28	33	7	13	20	26	33	39	8	16	24	31	39	47
1.6	5	9	14	19	23	28	6	11	17	22	28	33	7	13	20	27	33	40	8	16	24	32	40	48
1.8	5	10	15	19	24	29	6	11	17	23	28	34	7	14	21	27	34	41	8	16	25	33	41	49
2	5	10	15	19	24	29	6	12	13	23	29	35	7	14	21	27	34	41	8	17	25	33	42	50
2.2	5	10	15	20	25	30	6	12	18	23	29	35	7	14	21	28	35	42	9	17	26	34	43	51
2.4	5	10	15	20	25	30	6	12	19	24	30	36	7	14	22	29	36	43	9	17	26	35	43	52
2.6	5	10	16	21	26	31	6	12	19	25	31	37	7	15	22	29	37	44	9	18	27	35	44	53
2.8	5	10	16	21	26	31	6	12	19	25	31	37	8	15	23	30	38	45	9	18	27	36	45	54
3	5	11	16	21	27	32	6	13	19	25	32	38	8	15	23	31	38	46	9	18	28	37	46	55

CHLORINE CONCENTRATION (mg/L)	pH=8.0 Log Inactivation						pH=8.5 Log Inactivation						pH=9.0 Log Inactivation					
	0.5	1.0	1.5	2.0	2.5	3.0	0.5	1.0	1.5	2.0	2.5	3.0	0.5	1.0	1.5	2.0	2.5	3.0
<=0.4	8	17	25	33	42	50	10	20	30	39	49	59	12	23	35	47	58	70
0.6	9	17	26	34	43	51	10	20	31	41	51	61	12	24	37	49	61	73
0.8	9	18	27	35	44	53	11	21	32	42	53	63	13	25	38	50	63	75
1	9	19	27	36	45	54	11	22	33	43	54	65	13	26	39	52	65	78
1.2	9	18	28	37	46	55	11	22	34	45	56	67	13	27	40	53	67	80
1.4	10	19	29	38	48	57	12	23	35	46	58	69	14	27	41	55	68	82
1.6	10	19	29	39	48	58	12	23	35	47	58	70	14	28	42	56	70	84
1.8	10	20	30	40	50	60	12	24	36	48	60	72	14	29	43	57	72	86
2	10	20	31	41	51	61	12	25	37	49	62	74	15	29	44	59	73	89
2.2	10	21	31	41	52	62	13	25	38	50	63	75	15	30	45	60	75	90
2.4	11	21	32	42	53	63	13	26	39	51	64	77	15	31	46	61	77	92
2.6	11	22	33	43	54	65	13	26	39	52	65	78	16	31	47	63	78	94
2.8	11	22	33	44	55	66	13	27	40	53	67	80	16	32	48	64	80	96
3	11	22	34	45	56	67	14	27	41	54	68	81	16	32	49	65	81	97

CT Values for Inactivation of Viruses by Free Chlorine, pH 6.0–9.0

	Temperature (°C)																									
Inactivation (log)	0.5	1	2	3	4	5	6	7	8	9	10	11	12	13	14	15	16	17	18	19	20	21	22	23	24	25
2	6.0	5.8	5.3	4.9	4.4	4.0	3.8	3.6	3.4	3.2	3.0	2.8	2.6	2.4	2.2	2.0	1.8	1.6	1.4	1.2	1.0	1.0	1.0	1.0	1.0	1.0
3	9.0	8.7	8.0	7.3	6.7	6.0	5.6	5.2	4.8	4.4	4.0	3.8	3.6	3.4	3.2	3.0	2.8	2.6	2.4	2.2	2.0	1.8	1.6	1.4	1.2	1.0
4	12.0	11.6	10.7	9.8	8.9	8.0	7.6	7.2	6.8	6.4	6.0	5.6	5.2	4.8	4.4	4.0	3.8	3.6	3.4	3.2	3.0	2.8	2.6	2.4	2.2	2.0

Modified by linear interpolation between 5°C increments

CT Values for Inactivation of *Giardia* Cysts by Chlorine Dioxide, pH 6.0–9.0

	Temperature (°C)																								
Inactivation (log)	1	2	3	4	5	6	7	8	9	10	11	12	13	14	15	16	17	18	19	20	21	22	23	24	25
0.5	10.0	8.6	7.2	5.7	4.3	4.2	4.2	4.1	4.1	4.0	3.8	3.7	3.5	3.4	3.2	3.1	2.9	2.8	2.6	2.5	2.4	2.3	2.2	2.1	2.0
1	21.0	17.9	14.9	11.8	8.7	8.5	8.3	8.1	7.9	7.7	7.4	7.1	6.9	6.6	6.3	6.0	5.8	5.5	5.3	5.0	4.7	4.5	4.2	4.0	3.7
1.5	32.0	27.3	22.5	17.8	13.0	12.8	12.6	12.4	12.2	12.0	11.6	11.2	10.8	10.4	10.0	9.5	9.0	8.5	8.0	7.5	7.1	6.7	6.3	5.9	5.5
2	42.0	35.8	29.5	23.3	17.0	16.6	16.2	15.8	15.4	15.0	14.6	14.2	13.8	13.4	13.0	12.4	11.8	11.2	10.6	10.0	9.5	8.9	8.4	7.8	7.3
2.5	52.0	44.5	37.0	29.5	22.0	21.4	20.8	20.2	19.6	19.0	18.4	17.8	17.2	16.6	16.0	15.4	14.8	14.2	13.6	13.0	12.2	11.4	10.6	9.8	9.0
3	63.0	53.8	44.5	35.3	26.0	25.4	24.8	24.2	23.6	23.0	22.2	21.4	20.6	19.8	19.0	18.2	17.4	16.6	15.8	15.0	14.2	13.4	12.6	11.8	11.0

Modified by linear interpolation between 5°C increments.

CT Values for Inactivation of Viruses by Chlorine Dioxide, pH 6.0–9.0

	Temperature (°C)																								
Inactivation (log)	1	2	3	4	5	6	7	8	9	10	11	12	13	14	15	16	17	18	19	20	21	22	23	24	25
2	8.4	7.7	7.0	6.3	5.6	5.3	5.0	4.8	4.5	4.2	3.9	3.6	3.4	3.1	2.8	2.7	2.5	2.4	2.2	2.1	2.0	1.8	1.7	1.5	1.4
3	25.6	23.5	21.4	19.2	17.1	16.2	15.4	14.5	13.7	12.8	12.0	11.1	10.3	9.4	8.6	8.2	7.7	7.3	6.8	6.4	6.0	5.6	5.1	4.7	4.3
4	50.1	45.9	41.8	37.6	33.4	31.7	30.1	28.4	26.8	25.1	23.4	21.7	20.1	18.4	16.7	15.9	15.0	14.2	13.3	12.5	11.7	10.9	10.0	9.2	8.4

Modified by linear interpolation between 5°C increments.

CT Values for Inactivation of *Giardia* Cysts by Chloramine, pH 6.0–9.0

	Temperature (°C)																								
Inactivation (log)	**1**	**2**	**3**	**4**	**5**	**6**	**7**	**8**	**9**	**10**	**11**	**12**	**13**	**14**	**15**	**16**	**17**	**18**	**19**	**20**	**21**	**22**	**23**	**24**	**25**
0.5	635	568	500	433	365	354	343	332	321	310	298	286	274	262	250	237	224	211	198	185	173	161	149	137	125
1	1,270	1,136	1,003	869	735	711	687	663	639	615	592	569	546	523	500	474	448	422	396	370	346	322	298	274	250
1.5	1,900	1,700	1,500	1,300	1,100	1,066	1,032	998	964	930	894	858	822	786	750	710	670	630	590	550	515	480	445	410	375
2	2,535	2,269	2,003	1,736	1,470	1,422	1,374	1,326	1,278	1,230	1,184	1,138	1,092	1,046	1,000	947	894	841	788	735	688	641	594	547	500
2.5	3,170	2,835	2,500	2,165	1,830	1,772	1,714	1,656	1,598	1,540	1,482	1,424	1,366	1,308	1,250	1,183	1,116	1,049	982	915	857	799	741	683	625
3	3,800	3,400	3,000	2,600	2,200	2,130	2,060	1,990	1,920	1,850	1,780	1,710	1,640	1,570	1,500	1,420	1,340	1,260	1,180	1,100	1,030	960	890	820	750

Modified by linear interpolation between 5°C increments.

CT Values for Inactivation of Viruses by Chloramine

	Temperature (°C)																								
Inactivation (log)	**1**	**2**	**3**	**4**	**5**	**6**	**7**	**8**	**9**	**10**	**11**	**12**	**13**	**14**	**15**	**16**	**17**	**18**	**19**	**20**	**21**	**22**	**23**	**24**	**25**
2	1,243	1,147	1,050	954	857	814	771	729	686	643	600	557	514	471	428	407	385	364	342	321	300	278	257	235	214
3	2,063	1,903	1,743	1,583	1,423	1,352	1,281	1,209	1,138	1,067	996	925	854	783	712	676	641	605	570	534	498	463	427	392	356
4	2,883	2,659	2,436	2,212	1,988	1,889	1,789	1,690	1,590	1,491	1,392	1,292	1,193	1,093	994	944	895	845	796	746	696	646	597	547	497

Modified by linear interpolation between 5°C increments.

CT Values for Inactivation of *Giardia* Cysts by Ozone

	Temperature (°C)																								
Inactivation (log)	**1**	**2**	**3**	**4**	**5**	**6**	**7**	**8**	**9**	**10**	**11**	**12**	**13**	**14**	**15**	**16**	**17**	**18**	**19**	**20**	**21**	**22**	**23**	**24**	**25**
0.5	0.48	0.44	0.40	0.36	0.32	0.30	0.28	0.27	0.25	0.23	0.22	0.20	0.19	0.17	0.16	0.15	0.14	0.14	0.13	0.12	0.11	0.10	0.10	0.09	0.08
1.0	0.97	0.89	0.80	0.72	0.63	0.60	0.57	0.54	0.51	0.48	0.45	0.42	0.38	0.35	0.32	0.30	0.29	0.27	0.26	0.24	0.22	0.21	0.19	0.18	0.16
1.5	1.50	1.36	1.23	1.09	0.95	0.90	0.86	0.81	0.77	0.72	0.67	0.62	0.58	0.53	0.48	0.46	0.43	0.41	0.38	0.36	0.34	0.31	0.29	0.26	0.24
2.0	1.90	1.75	1.60	1.45	1.30	1.23	1.16	1.09	1.02	0.95	0.89	0.82	0.76	0.69	0.63	0.60	0.57	0.54	0.51	0.48	0.45	0.42	0.38	0.35	0.32
2.5	2.40	2.20	2.00	1.80	1.60	1.52	1.44	1.36	1.28	1.20	1.12	1.04	0.95	0.87	0.79	0.75	0.71	0.68	0.64	0.60	0.56	0.52	0.48	0.44	0.40
3.0	2.90	2.65	2.40	2.15	1.90	1.81	1.71	1.62	1.52	1.43	1.33	1.24	1.14	1.05	0.95	0.90	0.86	0.81	0.77	0.72	0.67	0.62	0.58	0.53	0.48

Modified by linear interpolation between 5°C increments.

CT Values for Inactivation of Viruses by Ozone

	Temperature (°C)																								
Inactivation (log)	**1**	**2**	**3**	**4**	**5**	**6**	**7**	**8**	**9**	**10**	**11**	**12**	**13**	**14**	**15**	**16**	**17**	**18**	**19**	**20**	**21**	**22**	**23**	**24**	**25**
2	0.90	0.83	0.75	0.68	0.60	0.58	0.56	0.54	0.52	0.50	0.46	0.42	0.38	0.34	0.30	0.29	0.28	0.27	0.26	0.25	0.23	0.21	0.19	0.17	0.15
3	1.40	1.28	1.15	1.03	0.90	0.88	0.86	0.84	0.82	0.80	0.74	0.68	0.62	0.56	0.50	0.48	0.46	0.44	0.42	0.40	0.37	0.34	0.31	0.28	0.25
4	1.80	1.65	1.50	1.35	1.20	1.16	1.12	1.08	1.04	1.00	0.92	0.84	0.76	0.68	0.60	0.58	0.56	0.54	0.52	0.50	0.46	0.42	0.38	0.34	0.30

Modified by linear interpolation between 5°C increments.

Appendix C

Chlorine Residual Test Methods

PURPOSE OF TEST

Chlorine is used primarily as a disinfectant to destroy disease-producing microorganisms in the water supply. In addition, it can improve the quality of finished water by reacting with ammonia, iron, manganese, sulfide, and some taste- and odor-producing substances.

Chlorine can also react with natural precursors producing undesirable disinfection by-products (DBPs) that may have adverse health effects. Excess chlorine also can cause taste and odor problems. Combined chlorine, formed when chlorine reacts with ammonia or amine compounds, can adversely affect aquatic life and persons with certain medical conditions.

Two types of chlorine residual are produced in water from the chlorination process: free available residual and combined available residual. *Free available residual* occurs when the water is thoroughly chlorinated. It exists in three forms: molecular chlorine (Cl_2), hypochlorous acid (HOCl), and hypochlorite ion (OCl^-). Molecular chlorine exists in the pH range of 1–4; hypochlorous acid in the pH range of 1–9 (it is the predominant form in the pH range of 2–7); hypochlorous acid and hypochlorite ion coexist in equal proportions at pH 7.4. Hypochlorite ion is the predominant form above pH 9.5. *Combined available residual* forms when chlorine reacts with ammonia that either occurs naturally or is added in treatment. It exists in three forms: monochloramine (NH_2Cl), dichloramine ($NHCl_2$), and trichloramine (NCl_3). Combined chlorine is a less effective disinfectant than free chlorine.

Chlorine residual analytical procedures measure free and total residuals, the sum of which is the total chlorine residual. Therefore, total chlorine residual is always equal to or greater than free chlorine residual.

LIST OF SIMPLIFIED METHODS

- Field method using commercial comparator kit
- Amperometric titration method
- Titrimetric method
- Colorimetric method

For other methods, refer to *Standard Methods for the Examination of Water and Wastewater,* Section 4500—Cl (APHA et al., latest edition):

- DPD colorimetric method
- Iodometric method
- Low-level amperometric titration method
- Free (available) chlorine test, syringaldazine (FACTS) method
- Iodometric electrode method

SIMPLIFIED PROCEDURES

Chlorine (Residual) Field Method Using Commercial Comparator Kit

This simple test is designed for field or laboratory determination of free and total chlorine in water using *N,N*-diethyl-*p*-phenylenediamine (DPD) as the color indicator. DPD produces a pink color to a degree proportional to the chlorine content of the water. The color of the water is then compared to the standard color scale in the comparator to determine the chlorine content of the water.

In this procedure, a sample is added to two identical viewing tubes. DPD is added to one tube to produce a color change. The tube without the indicator serves as a control tube to ensure an accurate color match by nullifying color or turbidity in the sample. Match the color of the sample containing indicator to the standard color scale (which overlays the control tube) and read the chlorine content directly from the comparator.

Warnings/cautions. The buffer pillow lowers the sample pH within the range of 6.2–6.5 for accurate results. A lower pH enables chloramine to appear as free chlorine. A higher pH causes dissolved oxygen to give a pink color identical to that produced by chlorine.

If the sample contains oxidized manganese, an inhibitor (included in the kit) must be used.

High temperature enables chloramine to appear as free chlorine and increases color fading. Complete the measurements rapidly at high temperatures.

Oxidized manganese that is naturally occurring or added during water treatment as potassium permanganate ($KMnO_4$) reacts with DPD to give a pink color identical to that produced by chlorine. In such a case a correction must be made for this interference. (See *Standard Methods* for further discussion.)

Chlorine dioxide (ClO_2), if present, appears with free chlorine to the extent of one-fifth the total chlorine content. (See *Standard Methods* for further discussion.)

Monochloramine (NH_2Cl), if present in high concentration, interferes in the free chlorine determination after 1 min of developing time. Therefore, all readings must be made within the specified time interval.

The DPD color comparator kit must be calibrated initially when the kit is purchased and at least once every 3 months thereafter. Color standards are light sensitive and fade when exposed to sunlight or high temperatures. Color comparators should be stored in a dark, cool location. Storing test kits inside vehicles shortens the life expectancy of color standards. Similarly, the shelf life of powder pillows or tablets is adversely affected when exposed to higher temperatures or direct sunlight. If the contents of the powder pillows or tablets are discolored, they should be discarded.

Apparatus.

- A color comparator kit complete with standard color scale (disk or blocks of appropriate chlorine range), color viewing tubes, and caps (commercially available)
- A 20- to 200-µL pipet
- A 1-L volumetric flask

Reagents.

DPD indicator. Commercially available as DPD Free Chlorine Reagent and DPD Total Chlorine Reagent; or prepare according to *Standard Methods*, Section 4500-Cl G, DPD Colorimetric Method.

Distilled water. Use reagent-grade, deionized distilled water, which should be chlorine free.

Potassium permanganate stock standard, $KMnO_4$. Commercially available or prepare as follows: Weigh 0.8910 g desiccated, reagent-grade potassium permanganate with an analytical balance. Transfer to a 1-L volumetric flask and bring to volume with distilled water. This solution is equivalent to 1,000 mg/L (ppm) of chlorine. Using volumetric glassware, dilute 10 mL of the 1,000 mg/L stock with distilled water to a final volume of 100 mL. (This solution must be made fresh for each calibration.) This solution is equivalent to 100 mg/L of chlorine.

Free chlorine determination procedure.

1. Rinse two viewing tubes with distilled water.
2. Fill a control viewing tube to the graduation mark with water to be tested and place it in the opening of the color comparator behind the standard color scale.
3. Fill sample viewing tube to graduation mark with water to be tested.
4. Add entire contents of prepackaged free chlorine reagent to sample viewing tube. Cap and swirl to mix. (Occasionally, swirling alone does not produce good precision in data. To solve this problem, shake the sample vigorously for 15 sec, then swirl to remove air bubbles. The powder does not have to dissolve completely to obtain a correct reading.) Place the sample tube in the comparator beside the control viewing tube. The free chlorine must be read in 1 min.
5. Hold the comparator to a light source and compare the sample viewing tube to the comparator standard color scale. When a color match is achieved, record the free chlorine value in milligrams per liter. The free chlorine must be read in 1 min.

Total chlorine determination procedure.

1. Rinse two viewing tubes with distilled water.
2. Fill the control viewing tube to the graduation mark with water to be tested and place in the opening of the color comparator behind the standard color scale.
3. Fill the sample viewing tube to the graduation mark with the water to be tested.

completely to obtain correct readings. Place the sample tube in the comparator beside the control viewing tube. Let stand at least 3 min but not more than 6 min.

5. Hold the comparator to a light source and compare the sample viewing tube to the comparator standard color scale. When a color match is achieved, record the total chlorine value in milligrams per liter.

Comparator calibration procedure. Known amounts of a potassium permanganate solution are used as standards to verify the calibration of the comparator standard color scale used to determine chlorine concentration. For more information on preparing a standard curve, see Standard Calibration Curves in chapter 1, AWWA Manual M12 (AWWA 2002).

Prepare standards as follows:

1. Fill a viewing tube to the graduation mark with distilled water.
2. Using a calibrated automatic microliter pipet, spike the tube with a known amount of potassium permanganate 100-mg/L chlorine-equivalent solution.
3. Prepare at least six standards that span the entire chlorine range of the comparator standard color scale being calibrated. Determine the true concentrations of the standards by using Table C-1.
   NOTE: Because of different viewing tube volumes, the resulting concentration will vary for a given potassium permanganate spike volume. Process the standards as described here and read the resulting color development against the comparator standard color scale. Compare the true value of the potassium permanganate standard with the observed value of the kit. Acceptable values are ±10 percent of the true value. If concentrations are outside the acceptable range, repair or replace the kit, calculate a correction factor, or draw a correction curve to determine sample concentration.

Table C-1 Calibrating comparator standard using potassium permanganate

$KMnO_4$ Spike Volume (μL)	Viewing Tube Volume			$KMnO_4$ Spike Volume (μL)	Viewing Tube Volume		
	5 mL	10 mL	25 mL		5 mL	10 mL	25 mL
5	0.10			90	1.80	0.90	
10	0.20	1.10		100	2.00	1.00	0.40
20	0.40	0.20		110	2.00	1.10	
25	0.50	0.25	0.10	125	2.50	1.25	0.50
30	0.60	0.30		140	2.80	1.40	
40	0.80	0.40		150	3.00	1.50	0.60
50	1.00	0.50	0.20	160	3.20	1.60	
60	1.20	0.60		170	3.40	1.70	
70	1.40	0.70		175	3.50	1.75	0.70
75	1.50	0.75	0.30	180	3.60	1.80	
80	1.60	0.80		190	3.80	1.90	
				200	4.00	2.00	0.80

Chlorine (Residual) Amperometric Method

This titration is designed primarily for laboratory rather than field use because it requires more skill and care than the colorimetric methods. Differentiation between free and combined chlorine is possible by pH adjustment and the presence or absence of potassium iodide (KI). Free chlorine can be determined at a pH between 6.5 and 7.5. Combined chlorine can be determined at a pH between 3.5 and 4.5 in the presence of the correct amount of potassium iodide.

Warnings/cautions. Phenylarsine oxide (C_6H_5AsO) titrant is a severe poison and a suspected carcinogen. Read all warnings regarding proper handling. Control of pH is important for correct results. At a pH above 7.5, the reaction with free chlorine becomes sluggish. At a pH below 6.5, some combined chlorine may react even in the absence of iodide. At a pH below 3.5, oxidized manganese reacts with the titrant. At a pH above 4.5, the titration for combined chlorine fails to reach completion.

High temperatures and prolonged titration time allow monochloramine to be titrated as free chlorine, which leads to an apparent increase in free chlorine results.

Chlorine dioxide and free halogens titrate as free chlorine, which leads to an apparent increase in free chlorine results. (See *Standard Methods* for further discussion.)

Excessive stirring of some commercial titrators can lower chlorine values by volatilization. Complete the analysis promptly.

Chlorine residuals higher than 2 mg/L are more accurately determined by using smaller sample volumes or by sample dilution.

NOTE: Perform the analysis immediately after collecting the sample.

Apparatus.

Amperometric titrator. A typical amperometric titrator consists of a two-electrode cell connected to a microammeter and an adjustable potentiometer. The cell unit includes a noble metal electrode, a reference electrode of silver-silver chloride in a saturated sodium chloride (NaCl) solution, and a salt bridge. An agitator and a buret are also required and are typically supplied with the titrator.

For best results, observe the following practices to prepare and operate the apparatus:

- Keep noble metal electrode free of deposits. Occasional mechanical cleaning with a suitable abrasive is sufficient.
- Keep the salt bridge in good operating condition. If plugging or improper flow of salt solution occurs in the salt bridge, empty the old material from the cell and replace it with fresh salt.
- Keep an adequate supply of solid salt in the reference electrode at all times.
- Thoroughly clean the agitator and exposed electrode system to remove the chlorine-consuming contaminants by immersion for several minutes in water containing 1–2 mg/L free available residual chlorine. Add potassium iodide to the same water and immerse for another 5 min.
- Thoroughly rinse the sensitized electrodes and agitator with distilled water or the sample to be tested.
- If the chlorine concentration of the samples approximates 0.5 mg/L, condition the electrode system further by conducting two or more titrations at the 0.5-mg/L level until the titrations become reproducible.
- Satisfy the chlorine demand of all glassware to be used for sampling and titrating samples by subjecting the critical surfaces to a water that contains 10 mg/L or more residual chlorine for at least 3 hr. Rinse with distilled water to remove the residual chlorine traces.

Reagents.

Phenylarsine oxide titrant, C_6H_5AsO (0.00564N) (PAO). Purchase this reagent, or prepare as follows: Dissolve approximately 0.8 g phenylarsine oxide powder in 150 mL 0.3*N* sodium hydroxide solution. After settling, decant 110 mL into 800 mL distilled water and mix thoroughly. Bring to a pH of 6–7 with 6*N* hydrochloric acid and dilute to 950 mL with distilled water.

Caution: Severe poison, suspected cancer agent

Phenylarsine oxide titrant standardization.

1. Place 200 mL distilled water in the titrating vessel and turn on the stirrer.
2. Add 1 mL of 20 percent sulfuric acid solution.
3. Add 1 mL potassium iodide solution.
4. Carefully add 5.0 mL 0.0025*N* potassium bi-iodate solution. A pale yellow color should develop.
5. Titrate to the usual amperometric end point (total chlorine).
6. Repeat the standardization process. Duplicate titrations should agree within 0.05 mL.

The normality (*N*) of the PAO titrant can be determined by the following formula:

N PAO × mL PAO = N potassium bi-iodate × mL potassium bi-iodate

$$N\text{PAO} = \frac{0.0025N \times 5 \text{ mL}}{\text{mL PAO used}} \quad \text{(C-1)}$$

The acceptable range of the PAO titrant normality is ±5 percent of theoretical normality (0.00564*N*). Acceptable range = 0.005358*N* to 0.005922*N*. If the normality of the titrant is outside the acceptable range, replace the titrant.

Phosphate buffer solution, pH 7.

1. Dissolve 25.4 g anhydrous potassium dihydrogen phosphate (also called potassium monobasic phosphate [KH_2PO_4]) and 34.1 g anhydrous disodium hydrogen phosphate (also called sodium dibasic phosphate [Na_2HPO_4]) in 800 mL distilled water.
2. Add 2 mL sodium hypochlorite solution that contains 5 percent available chlorine (common household bleach). Stopper and mix thoroughly.
3. Store in a cool, dark place away from sunlight or heat for two days, so the chlorine can react completely with the ammonium contaminants usually present in phosphates.
4. Place the bottle in sunlight, indoors or outdoors, until all of the chlorine disappears. The time required will vary from one day during summer to one week during winter.
5. When no total chlorine remains, transfer the contents of the bottle to a 1-L graduated cylinder and dilute to the 1-L mark with distilled water. Mix thoroughly by pouring back into the bottle.
6. Filter the solution if any precipitate forms on standing.

Potassium iodide solution.

1. Place 105 mL distilled water in a 250-mL flask and boil for 7–10 min. Cover the top of the flask with a clean, small, inverted beaker and allow the water to cool to room temperature. To hasten cooling, place the flask in a bath of cold running water.

2. On a rough balance, weigh 5 g potassium iodide. Transfer to the freshly boiled and cooled distilled water and mix thoroughly.
3. Transfer the solution to an amber, glass-stoppered bottle. Store in a dark, cool place, preferably a refrigerator. Discard the solution when a yellow color develops.

Acetate buffer solution, pH 4.

1. Measure 400 mL distilled water into a 1.5-L beaker. Prepare buffer solution in a fume hood.
2. With a 1-L graduated cylinder, measure 480 mL concentrated acetic acid (CH_3COOH) (also called glacial acetic acid) and add to the 400 mL distilled water. Mix thoroughly.
3. Weigh 243 g sodium acetate trihydrate ($NaC_2H_3O_2 \cdot 3H_2O$) and dissolve in the acetic acid solution.
4. Transfer the solution to the 1-L graduated cylinder; dilute to the 1-L mark with distilled water and mix thoroughly.

Distilled water. Use reagent-grade, deionized distilled water, which should be chlorine free.

Potassium bi-iodate, $KH(IO_3)_2$. This reagent is available commercially from major distributors at 0.025*N* and must be diluted using volumetric glassware to 0.0025*N*. Dilute 10 mL 0.025*N* potassium bi-iodate with distilled water to a final volume of 100 mL. The solution must be made fresh for each standardization.

1. To prepare the reagent in the laboratory, dry 2–4 g of reagent-grade potassium bi-iodate for 2 hr at 105°C (221°F).
2. Desiccate to room temperature.
3. Dissolve 1.6245 g potassium bi-iodate in distilled water. Using a 500-mL volumetric flask, dilute to a final volume of 500 mL. This is a 0.1*N* solution.
4. Using volumetric glassware, dilute 25 mL 0.1*N* solution with distilled water to a final volume of 1 L. This is a 0.0025*N* solution and must be made fresh for each standardization.

Sulfuric acid, H_2SO_4, 20 percent solution.

Free chlorine determination procedure.

1. Fill the buret with phenylarsine oxide titrant. Record the liquid level in the buret by reading at the bottom of the meniscus. Guard against a leaky stopcock, which can result in the loss of titrant.
2. Select the sample volume. Measure the sample and distilled water volumes for the indicated residual chlorine ranges.

   If the residual chlorine falls within the range of 0.0 to 4.0 mg/L, use 100 mL sample and distilled water. First, mix 5 mL phosphate buffer solution and 5 mL DPD reagent in a 250-mL flask. Add 100 mL sample and mix. (If the sample is added before buffer, the test does not work.)

   If the residual chlorine exceeds 4.0 mg/L, reduce the sample size and dilute the sample with distilled water to a total volume of 100 mL (Table C-2). First, mix 5 mL phosphate buffer solution and 5 mL DPD reagent in a 250-mL flask. Add the specified volume of chlorine free water. Add the sample and mix.
3. Unless sample pH is known to be between 6.5 and 7.5, add 1 mL pH 7 phosphate buffer solution.

Table C-2 Dilution table for various strengths of residual chlorine

Residual Chlorine Range (mg/L)	Original Sample Volume (mL)	Distilled Water Volume (mL)
0.0–4.0	100	0
4.1–8.0	50	50
8.1–16.0	25	75

Table C-3 Calculating free available chlorine from amperometric titration results

Sample Volume (mL)	Multiply Milligram per Liter Titrant Used By
200	1
100	2
50	4

4. Titrate with standard phenylarsine oxide titrant, watching the needle movement on the microammeter scale. When the needle moves to the 0 end of the scale, return to mid-scale with the proper adjustment for easier observation and greater sensitivity. As the needle activity diminishes, add progressively smaller increments of titrant. Make successive buret readings when the needle action becomes sluggish, signaling the approach of the end point. Subtract the last very small increment that causes no needle response because of overtitration.
5. Read the new buret level at the bottom of the meniscus and calculate the volume of titrant used by subtracting the initial buret reading (step 1) from the end point reading.
6. Calculate free available chlorine by multiplying the result found in step 5 by the appropriate factor (Table C-3). (Use distilled water to bring the sample volume to 200 mL.)

Combined chlorine determination procedure.

7. To the sample remaining from free chlorine titration, add exactly 1 mL potassium iodide solution.
8. Add 1 mL acetate buffer solution to the sample.
9. Do not refill buret, but continue titration after recording figure for free chlorine. Repeat the titration procedure described in step 4.
10. Read the new buret level at the bottom of the meniscus and record the total volume of titrant used in both the free available chlorine titration and the combined available chlorine titration. This figure represents total chlorine. Multiply this total by the factor given in step 6.
11. Subtract the value in step 6 (free available chlorine) from the value in step 10 (total chlorine) to obtain the combined available chlorine.

    Caution: Wash the electrodes, stirrer, and sample container thoroughly to remove every trace of iodide from the apparatus before making the next free available chlorine determination. Confirm complete iodide removal by duplicating the sample.

Chlorine (Residual) Titrimetric Method

N,N-diethyl-*p*-phenylenediamine (DPD) is used as an indicator in the titrimetric procedure with ferrous ammonium sulfate ($Fe[NH_4]_2[SO_4]_2 \cdot 6H_2O$). This simplified procedure is used to determine free, combined, or total chlorine.

Warnings/cautions. Sample pH must be between 6.2 and 6.5 for accurate results. A lower pH enables chloramine to appear as free chlorine. A higher pH causes dissolved oxygen to give a pink color identical to that produced by chlorine. High temperature enables chloramine to appear as free chlorine and increases color fading. Complete the measurements rapidly at high temperatures.

Oxidized manganese that is naturally occurring or added during water treatment as potassium permanganate reacts with DPD to give a pink color identical to that produced by chlorine. In such a case, a correction must be made for this interference. (See *Standard Methods* for further discussion.)

Chlorine dioxide, if present, appears with free chlorine to the extent of one-fifth the total chlorine content. (See *Standard Methods* for further discussion.)

Monochloramine, if present in high concentration, interferes in the free chlorine determination after 1 min of developing time. Therefore, all readings must be made within the specified time interval.

Apparatus.

- 10-mL buret and support
- 100-mL graduated cylinder or volumetric pipets for measuring the sample
- One or more 250-mL flasks
- Two 5-mL pipets for dispensing DPD reagent and phosphate buffer solution
- Spatula for dispensing potassium iodide crystals
- Dropping pipet or medicine dropper for dispensing sodium arsenite (also called sodium meta-arsenite [$NaAsO_2$]) solution

Reagents.

Distilled water. Use reagent-grade, deionized distilled water, which should be chlorine free.

Dilute sulfuric acid solution. Using a 50-mL graduated cylinder, measure 30 mL distilled water and pour into a 100-mL beaker.

1. With a 10-mL pipet, measure 10 mL concentrated sulfuric acid, H_2SO_4.
2. While stirring, slowly and cautiously add 10 mL sulfuric acid to 30 mL distilled water. Considerable heat is generated by mixing acid and water, so pour slowly and mix well to avoid dangerous spattering. Cool to room temperature before use.

Ferrous ammonium sulfate titrant.

1. Measure 1,200 mL distilled water into a 2-L flask and boil for 5 min. Cover the top of the flask with an inverted, 400-mL beaker and allow to cool to room temperature. Place in a cold water bath to speed cooling.
2. Pour half the freshly boiled and cooled distilled water into a 1.5-L beaker. Add 1 mL dilute sulfuric acid solution and mix.
3. On an analytical balance, weigh 1.106 g reagent-grade ferrous ammonium sulfate ($Fe[NH_4]_2[SO_4]_2 \cdot 6H_2O$). Carefully transfer to the 1.5-L beaker and dissolve in the distilled water prepared in step 2.
4. Transfer the solution to a 1-L volumetric flask. Rinse the beaker three times with 100-mL portions of distilled water. Add rinsings to flask. Dilute

to the mark with distilled water, stopper, and mix thoroughly. Note that 1 mL of this titrant is equivalent to 1 mg/L of chlorine in the titration procedure.

5. Store the titrant in an amber, glass-stoppered bottle away from bright light. Discard after 1 month.

N,N-*diethyl*-p-*phenylenediamine reagent.*

1. Place 600 mL distilled water in a 1.5-L beaker. Add 8 mL dilute sulfuric acid solution and mix.
2. Weigh 0.2 g disodium ethylenediaminetetraacetate dihydrate (also called ethylenedinitrilotetraacetic acid sodium salt [EDTA]). Carefully transfer to the 1.5-L beaker and dissolve by mixing in solution described in step 1.
3. Weigh either 1 g *N,N*-diethyl-*p*-phenylenediamine oxalate or 1.5 g p-amino-N, N-diethylaniline sulfate. Carefully transfer to the 1.5-L beaker and dissolve by mixing in solution described in step 2.
4. Transfer the combined solution (step 3) to a 1-L graduated cylinder and dilute to the 1-L mark with distilled water. Mix thoroughly by pouring back into the beaker and stirring.
5. Store the reagent solution in an amber, glass-stoppered bottle away from bright light. Discard when the solution becomes discolored.
   Caution: The oxalate reagent is toxic. Do not ingest. Dispense the solution with an automatic, safety, or bulb-operated pipet.

Phosphate buffer solution.

1. Weigh the following dry chemicals separately: (1) 24 g disodium hydrogen phosphate (also called sodium dibasic phosphate [Na_2HPO_4]) and (2) 46 g potassium dihydrogen phosphate (also called potassium monobasic phosphate [KH_2PO_4]).
2. Transfer the weighed chemicals to a 1.5-L beaker and dissolve in 600 mL distilled water. If necessary, heat the solution gently and stir to bring all the chemicals into solution. If heat is used to dissolve the chemicals, cool the solution to room temperature.
3. Weigh 0.8 g EDTA and dissolve in 100 mL distilled water. Add to solution (step 2) and mix.
4. Transfer the mixed solution (step 3) to a 1-L graduated cylinder and dilute to the 1-L mark with distilled water. Mix thoroughly by pouring back into the beaker and stirring.
5. Weigh 20 mg mercuric chloride ($HgCl_2$) and add to solution (step 4) to prevent mold growth and interference in the free available chlorine test caused by any trace of iodide in the reagents.
   Caution: Mercuric chloride is toxic. Do not ingest.
6. Potassium iodide crystals.
7. Sodium arsenite solution (to estimate manganese interference). Weigh 5 g sodium arsenite (also called sodium meta-arsenite [$NaAsO_2$]). Dissolve in 1 L distilled water.
   Caution: Poison. Handle with extreme caution. Do not ingest. Dispense the solution with an automatic, safety, or bulb-operated pipet.

Free available chlorine determination procedure.

1. Fill the buret with ferrous ammonium sulfate (FAS) ($Fe[NH_4]_2[SO_4]_2 \cdot 6H_2O$). Record the liquid level in the buret by reading the bottom of the meniscus. Guard against a leaky stopcock, which can result in the loss of titrant.
2. Select the sample volume. Measure the sample and distilled water volumes for the indicated residual chlorine ranges.

   If the residual chlorine falls within the range of 0.0–4.0 mg/L, use 100 mL sample and no distilled water. First, mix 5 mL phosphate buffer solution and 5 mL DPD reagent in a 250-mL flask. Add 100 mL sample and mix. (If the sample is added before buffer, the test does not work.)

   If the residual chlorine exceeds 4.0 mg/L, reduce the sample size and dilute the sample with distilled water to a total volume of 100 mL (Table C-2). First, mix 5 mL phosphate buffer solution and 5 mL DPD reagent in a 250-mL flask. Add the specified volume of distilled water. Add the sample and mix.
3. If the sample turns pink or red, add FAS titrant from the buret and swirl the flask constantly until the pink just disappears.
4. Read the new buret level at the bottom of the meniscus and calculate the volume of titrant used by subtracting the initial buret reading (step 1) from the present reading.
5. Calculate the free available chlorine by multiplying the result found in step 4 by the appropriate factor (see Table C-4).

Combined chlorine determination procedure.

6. Add several potassium iodide crystals (total weight 0.5–1.9 g) to the flask from step 5 (free available chlorine) and mix to dissolve. Let the solution stand for 2 min so that the chloramine in the sample can convert the iodide to iodine as evidenced by the return of the pink or red color.
7. Resume titrating with small volumes of FAS titrant until the pink or red color again disappears.
8. Read the new buret level at the bottom of the meniscus and record the total volume of titrant used in both the free available chlorine titration (step 4) and the combined chlorine titration (step 7). Multiply this total by the factor given in step 5.
9. Subtract the value in step 5 from the value in step 8 (combined chlorine) to obtain the value for combined available chlorine.

Table C-4 Calculating free available chlorine from titration results

Sample Volume (mL)	Multiply Milliliters of Titrant Used By
100	1
50	2
25	4

Estimating manganese interference procedure.

1. Place 5 mL phosphate buffer solution, one small crystal of potassium iodide, and 0.5 mL sodium arsenite solution in a 250-mL flask and mix.
2. Add 100 mL sample and mix.
3. Add 5 mL DPD reagent and mix.
4. If the solution turns pink or red, manganese interference is present. Titrate with FAS titrant until the pink disappears.
5. Read the new buret level at the bottom of the meniscus and calculate the volume of titrant used by subtracting the initial buret reading from the present reading. Multiply the result by the factor given in step 5 to obtain the manganese interference.
6. Subtract manganese interference from the results in step 5 (free chlorine) and step 8 (combined chlorine) to obtain the true, free available chlorine and total available chlorine, respectively.
7. If the values on the two spiked sample portions are higher by the amount that was artificially added, the result on the original unspiked sample can be assumed to be correct. If the recoveries exceed or fall short of the calculated amount by more than experimental error, the trouble may be attributed to an interference in the unknown sample.

REFERENCES

APHA, AWWA, and WEF (American Public Health Association, American Water Works Association, and Water Environment Federation). *Standard Methods for the Examination of Water and Wastewater.* Latest edition. Washington, D.C.: APHA.

American Water Works Association (AWWA). 2002. Manual M12, *Simplified Procedures for Water Examination.* Denver, Colo.: American Water Works Association.

Appendix D

Disinfection (Chlorination) of Facilities

Before any water treatment system can be put into service, the system must be disinfected and prepared to receive the water. The American Water Works Association (AWWA) has developed a series of standards covering disinfection from the well to the treatment plant and from storage facilities to the distribution system. Updated regularly, these standards present accepted procedures and practices to be taken prior to placing the system, tank, and piping system into service. Some of these standards are listed as follows:

AWWA C651: *Standard for Disinfecting Water Mains*
AWWA C652: *Standard for Disinfection of Water-Storage Facilities*
AWWA C653: *Standard for Disinfection of Water Treatment Plants*
AWWA C654: *Standard for Disinfection of Wells*

Important points taken from these standards are outlined here:

- The initial disinfection efforts should be directed toward cleaning the facilities. Pipes should be stored to prevent contamination prior to use and protected after installation. This is important because the pipe is not easily inspected once installed. Tanks, wells, and plant facilities should be thoroughly cleaned by washing with sufficient water to remove debris and dirt. Protect equipment after installation to reduce the possibility of contamination.
- Regardless of the facilities to be disinfected, the type of disinfectant used is usually chlorine gas, liquid sodium hypochlorite, or calcium hypochlorite tablets. The injection of either gas chlorine or sodium hypochlorite into the flowing water provides a more uniform-strength solution.
- Surface washing of tanks and other plant facilities with solutions of 200 ppm (mg/L) of hypochlorite is frequently practiced. Whether surface washing is practiced or the walls are simply flushed with high volumes of water, the facility must be exposed to a minimum concentration of chlorine residual for a fixed time period.

- Water mains require a 24-hr exposure with a minimum of 25 ppm (mg/L) of free chlorine, while tanks can be exposed to concentrations of 10 ppm (mg/L) of free chlorine with a concentration of 2 ppm (mg/L) at the end of 24 hr. Treatment plant disinfection requirements are similar to the water main and storage tank levels, while wells require a 50-ppm (mg/L) free chlorine exposure for 12–24 hr.
- Disposal of the highly chlorinated waters at the end of the disinfection period must follow and meet the appropriate regulations governing effluent discharge. The use of dechlorination agents, discharge into the local sewer system, or dilution prior to discharge may be satisfactory methods of disposal.
- Most state departments of health will provide guidelines for exposure time and minimum concentrations. The regulations of the state in which the utility is located must be known and followed.
- The residuals to be measured in these disinfection procedures are considerably higher than the operating levels at the treatment plant. Manual sampling of these levels and use of the drop dilution method to determine the residual level is necessary. The drop dilution method is covered in appendix C.

AWWA Manuals

M1, *Principles of Water Rates, Fees, and Charges,* Fifth Edition, 2000, #30001PA

M2, *Instrumentation and Control,* Third Edition, 2001, #30002PA

M3, *Safety Practices for Water Utilities,* Sixth Edition, 2002, #30003PA

M4, *Water Fluoridation Principles and Practices,* Fifth Edition, 2004, #30004PA

M5, *Water Utility Management Practices,* First Edition, 2005, #30005PA

M6, *Water Meters—Selection, Installation, Testing, and Maintenance,* Fourth Edition, 1999, #30006PA

M7, *Problem Organisms in Water: Identification and Treatment,* Third Edition, 2004, #30007PA

M9, *Concrete Pressure Pipe,* Second Edition, 1995, #30009PA

M11, *Steel Pipe—A Guide for Design and Installation,* Fifth Edition, 2004, #30011PA

M12, *Simplified Procedures for Water Examination,* Third Edition, 2002, #30012PA

M14, *Recommended Practice for Backflow Prevention and Cross-Connection Control,* Third Edition, 2003, #30014PA

M17, *Installation, Field Testing, and Maintenance of Fire Hydrants,* Third Edition, 1989, #30017PA

M19, *Emergency Planning for Water Utility Management,* Fourth Edition, 2001, #30019PA

M21, *Groundwater,* Third Edition, 2003, #30021PA

M22, *Sizing Water Service Lines and Meters,* Second Edition, 2004, #30022PA

M23, *PVC Pipe—Design and Installation,* Second Edition, 2002, #30023PA

M24, *Dual Water Systems,* Second Edition, 1994, #30024PA

M25, *Flexible-Membrane Covers and Linings for Potable-Water Reservoirs,* Third Edition, 2000, #30025PA

M27, *External Corrosion Introduction to Chemistry and Control,* Second Edition, 2004, #30027PA

M28, *Rehabilitation of Water Mains,* Second Edition, 2001, #30028PA

M29, *Water Utility Capital Financing,* Second Edition, 1998, #30029PA

M30, *Precoat Filtration,* Second Edition, 1995, #30030PA

M31, *Distribution System Requirements for Fire Protection,* Third Edition, 1998, #30031PA

M32, *Distribution Network Analysis for Water Utilities,* Second Edition, 2005, #30032PA

M33, *Flowmeters in Water Supply,* First Edition, 1989, #30033PA

M36, *Water Audits and Leak Detection,* Second Edition, 1999, #30036PA

M37, *Operational Control of Coagulation and Filtration Processes,* Second Edition, 2000, #30037PA

M38, *Electrodialysis and Electrodialysis Reversal,* First Edition, 1995, #30038PA

M41, *Ductile-Iron Pipe and Fittings,* Second Edition, 2003, #30041PA

M42, *Steel Water-Storage Tanks,* First Edition, 1998, #30042PA

M44, *Distribution Valves: Selection, Installation, Field Testing, and Maintenance,* First Edition, 1996, #30044PA

M45, *Fiberglass Pipe Design,* Second Edition, 2005, #30045PA

M46, *Reverse Osmosis and Nanofiltration,* First Edition, 1999, #30046PA

M47, *Construction Contract Administration,* First Edition, 1996, #30047PA

M48, *Waterborne Pathogens,* Second Edition, 2006, #30048PA

M49, *Butterfly Valves: Torque, Head Loss, and Cavitation Analysis,* First Edition, 2001, #30049PA

M50, *Water Resources Planning,* First Edition, 2001, #30050PA

To order any of these manuals or other AWWA publications, call the Bookstore toll-free at 1-800-926-7337.

M51, *Air-release, Air/Vacuum and Combination Air Valves,* First Edition, 2001, #30051PA

M52, *Water Conservation Programs—A Planning Manual,* First Edition, 2006, #30052PA

M53, *Microfiltration and Ultrafiltration Membranes for Drinking Water*, First Edition, 2005, #30053PA

M54, *Developing Rates for Small Systems,* First Edition, 2004, #30054PA

M55, *PE Pipe—Design and Installation,* First Edition, 2006, #30055PA

M56, *Fundamental and Control of Nitrification in Chloraminated Drinking Water Distribution Systems*, First Edition, 2006, #30056PA

Index

NOTE: *f.* indicates a figure; *t.* indicates a table.

This page intentionally blank.